IEE CONTROL ENGINEERING SERIES 13

SERIES EDITORS: G. A. MONTGOMERIE
PROF. H. NICHOLSON

MODELLING
OF DYNAMICAL
SYSTEMS Vol.2

Previous volumes in this series:

MODELLING
OF DYNAMICAL
SYSTEMS Vol. 2

Edited by
H. Nicholson, D.Eng., M.A., F.I.E.E., M.I.Mech.E.
Professor of Control Engineering
University of Sheffield
England

PETER PEREGRINUS LTD
on behalf of the
Institution of Electrical Engineers

Published by: The Institution of Electrical Engineers, London
and New York
Peter Peregrinus Ltd., Stevenage, UK, and New York

British Library Cataloguing in Publication Data

Modelling of dynamical systems.
    Vol. 2.—(Institution of Electrical Engineers.
    IEE control engineering series; vol. 13).
    1. System analysis
    2. Mathematical models
    I. Nicholson, Harold   II. Series
    511'.8   QA402   80-40463
ISBN 0-906048-45-1

Typeset by Santype International Limited, Salisbury
Printed in England by A. Wheaton & Co., Ltd., Exeter

Contents

Preface

In this Volume 2 of the book 'Modelling of Dynamical Systems', Chapters 1 to 4 represent a significant overall contribution devoted to ironmaking and steelmaking and steel processes. Chapter 1 discusses the modelling of agglomeration processes with particular emphasis on sintering and sinter strand control. The theme is continued in Chapter 2 with the development of a simple mathematical model for the argon-oxygen vessel used in steel refining. Chapter 3 then introduces various aspects of electric-arc furnace operation, including the modelling of electrode position controllers, high-power arc discharges and arc furnace supply systems. It also illustrates the application of system identification to supplement analytical modelling and to verify modelling assumptions. The design of a shape-control system for a single-strand cold-rolling mill which incorporates a static and dynamic model of a Sendzimir-type mill completes the steel processes section in Chapter 4.

In Chapter 5, the emphasis is on the representation and control of turbine-generator units and includes the development of physically realizable multi-variable controllers for turbogenerators based on nonlinear and linearized plant models. Chapter 6 then discusses the various approximations and solution techniques that have been used for the modelling and simulation of dynamic and steady-state flow in large-scale gas supply systems.

Chapter 7 investigates coal and mineral extraction processes and integrates the dynamics of the cutting process with the haulage and feed transmission system. The steering dynamics are also investigated and the chapter is completed with a proposed approach to a control design for the overall mining process. The various activities and communication flows required for the control of integrated production processes are discussed finally in Chapter 8, and illustrated by means of a structured analysis model for a batch manufacturing process.

The chapters of the book include a relatively wide range of modelling applications that highlight the methods and techniques which have been used to construct realistic system models based on physical laws and sound engineering judgment. It will be evident from the various contributions that the modelling of complex dynamical systems is often a difficult exercise, particularly when combined with data fitting and parameter estimation, and it is hoped that the work will provide some guidance and also illustrate the underlying principles required in mathematical system modelling.

H. NICHOLSON
University of Sheffield

July, 1980

List of contributors

S. A. BILLINGS

Department of Control Engineering, University of Sheffield, Mappin Street, Sheffield S1 3JD, England

T. R. CROSSLEY

Department of Mechanical Engineering, University of Salford, Salford M5 4WT, England

J. B. EDWARDS

Department of Control Engineering, University of Sheffield, Mappin Street, Sheffield S1 3JD, England

K. V. FERNANDO

Department of Control Engineering, University of Sheffield, Mappin Street, Sheffield S1 3JD, England

T. FINCHAM

British Gas Corporation, London Research Station, Michael Road, London SW6 2AD

M. H. GOLDWATER

British Gas Corporation, London Research Station, Michael Road, London SW6 2AD

M. J. GRIMBLE

Department of Electrical & Electronic Engineering, Sheffield City Polytechnic, Pond Street, Sheffield S1 1WB, England

B. HOGG

Department of Electrical and Electronic Engineering, The Queen's University of Belfast, Ashby Building, Stranmillis Road, Belfast BT9 5AH, Northern Ireland

H. NICHOLSON

Department of Control Engineering, University of Sheffield, Mappin Street, Sheffield S1 3JD, Sheffield, England

E. ROSE

Department of Control Engineering, University of Sheffield, Mappin Street, Sheffield S1 3JD, Sheffield, England

Chapter 1

Ironmaking and steelmaking—I

E. Rose

List of principal symbols

a	specific surface area
b	growth constant
c_g	specific heat of gas
c_p	specific heat of solid
c_g'	derivative of $c_g \tau$
c_p'	derivative of $c_p t$
C^*	indicates concentration at N.T.P.
$C_{O_2}, C_{CO_2}, C_{N_2}$	concentration of gas components
d_p	effective pellet diameter
D	diffusion coefficient
D_{ON}	diffusion coefficient of oxygen through nitrogen
$dP, \Delta P$	pressure drop
F_u	heat available for fusion
F/A	volumetric gas flow per unit area
G	mass velocity of gas
h	thickness of bed
h_c	thickness of reaction zone
h_p	convective heat transfer coefficient between solid and gas
H_c	heat of combustion in coke
H_v	latent heat of vaporization
K_c	chemical combustion rate constant
K_m	coefficient of mass transfer
l	length of bed
L_R	heat needed for limestone reduction
n	number of zones within a section of bed
n_c	number of coke particles per unit volume

N	total number of bed sections
N_u, R_e, P_l	Nusselt, Reynolds and Prandtl numbers
P	permeability
P_c	permeability of reaction zone
P_r	permeability of raw mix
r_c	effective radius of coke particles
R_c	combustion rate
s	suction
s_c	suction across reaction zone
s_{h-z}	suction across unprocessed zone
t	temperature of solid
T_1, T_2	time constants
u_j	heat-wave velocity, jth section
U	gas velocity $(= F/A)$
v	horizontal velocity of strand
v_a	horizontal strand velocity obtained by averaging
v_c	constant value of strand speed (gives correct burnthrough for midvalue of pellet diameter)
v_j	required horizontal velocity for jth section
V	specific volume of gas $1/\rho_g$
W	humidity of gas
W_G	mass of water condensed from gas
x_j	horizontal distance to burnthrough, jth section
X_w	mass of water in drying region
z	depth of leading edge of reaction zone
α'	shape factor
α, β	constants obtained from test pot data
ϵ	voidage
δz	thickness of band of gas
Δz_i	thickness of ith zone
$\Delta \phi$	update period for solid temperature
ρ_g	density of gas
ρ_p	density of solid
τ	temperature of gas
μ	viscosity of the gas
λ	thermal conductivity of the gas
θ	time
θ_{bj}	time to burnthrough, jth section
θ_r	section residence time

1.1 Agglomeration processes

Most of the world's iron is produced by the blast furnace process. It is therefore important that the blast furnace should operate efficiently and this will only occur if a large proportion of the feed to the furnace has been properly agglomerated. The main agglomeration processes are pelletizing and sintering, the products of which are often combined, together with raw ore and flux to form the blast furnace burden. Practices in different parts of the world vary according to the quality of ore available and the geographical location of the ore mines. During the past 25 years there has been much discussion of the relative merits of pelletizing and sintering typified perhaps by the title of Visvanathan's paper[30] 'Sinter versus Pellets'. Each method has its advocates but generally the two processes have come to be recognized as complementary in providing the right feed for the blast furnace. Ogg and Jennings[18] show how the world production of iron, sinter and pellets has developed since 1957 and in considering the economics of sintering emphasize the importance of regarding the process in the context of the total cost of ironmaking.

1.1.1 Pelletizing

The production of pellets is a two-stage process often carried out at the mine. Finely ground ore is mixed with a critical quantity of water and rolled in a drum to produce spheres typically 12 mm in diameter. A binder such as bentonite is included in the mix to hold the particles together and provide adequate strength to withstand handling. The so-called 'green balls' are then fed to an indurating machine, a feature of which is a continuously moving grate onto which the green balls are loaded. A stream of hot air drawn through the grate dries the pellets, they are then fired at a temperature between 1200°C and 1300°C and as they are further conveyed by the machine, cool air is drawn through the bed thereby reducing the temperature of the pellets. The heat transferred from the pellets raises the temperature of the air which is then recirculated for use in the drying phase of the process.

Pelletizing is a concentration process producing pellets which are rich in iron and are generally acidic, although much attention has recently been focused on the effects of adding excess flux at the balling stage.

1.1.2 Sintering

Sintering is mainly carried out by a continuous process using a machine based on that originally designed at the beginning of the twentieth century by Dwight and Lloyd for the sintering of copper ore. The sinter machine is normally situated adjacent to the blast furnace forming an integral part of the iron-producing plant. A wide range of quality of ore may be used in sintering.

A raw mix is formed from iron ore fines, coke and water and loaded onto a moving grate where it is levelled to form a uniform bed. The coke particles near the surface of the bed are ignited and as the material is conveyed along the 'strand' a heat wave progresses downwards as air is sucked into the surface of the bed and exhausted through windboxes situated below the strand.

During the process the volatiles are driven off, the hot material fuses at approximately 1400°C and as air is drawn in behind the fusion zone the material cools to form a friable cake of sinter which is unloaded from the strand and is broken down into small pieces from which fine material (−5 mm) is extracted by sieving before it is passed forward for use as blast furnace burden.

1.1.3 Requirements of blast furnace burden

Pelletizing and sintering are aimed at producing a good quality feed to the blast furnace which will facilitate uniform gas-solid contact across the stack, be chemically reducible and will keep the thermal demand on the blast furnace as low as possible.

High-grade ores may be fed directly to the blast furnace but the lump ore must first be broken down into a reasonably uniform size range (10–25 mm). Fines produced in breaking down the lump ore are sieved out and sintered.

Low-grade ores may be concentrated and pelletized or they may be sintered in order to make them acceptable as blast furnace burden.

In normal operating practice the liquid slag produced in the blast furnace is required to have a prescribed basicity (usually unity) and in order to achieve this value it is often necessary to add a flux, e.g. limestone, to the blast furnace burden. Alternatively, or additionally, flux may be added to the raw mix during pelletizing and sintering. The production of fluxed pellets having the required strength to withstand handling is particularly difficult and several investigations have recently been undertaken on this topic, e.g. Efimenko et al.[10]

On the other hand sinter is commonly fluxed to the extent that little or no addition of limestone to the furnace charge is required.

1.1.4 Modelling and simulation of agglomeration processes

The recent increase in computing power and the education of research engineers in control systems technology has led to the formulation and simulation of mathematical models representing particularly the pellet induration and sinter strand processes. The models are founded on fundamental theoretical studies and on the findings of pilot plant trials. Data obtained from sinter test pot investigations have been used extensively to provide numerical values for the model parameters and the validation of the model at various stages in its development.

The modelling of the pellet indurating process and the sinter strand involve the use of similar equations and simulation methods. Both are moving-grate processes in which heat is transferred between the bed of material loaded onto the grate and the air which is drawn through it. In the following pages attention is given to the dynamic modelling of the sinter strand and the reader is referred to Ball et al.,[1] Voskamp and Brasz[32] and Young[33] for information relating specifically to pelletizing.

1.2 An empirical model of the sinter strand process

A schematic diagram of a sinter strand is shown in Fig. 1.1. The raw mix, usually in the form of micropellets (ideally not less than 3 mm in diameter) is loaded evenly onto the grate which has first been covered by a hearth layer of presintered material. The pellets which have a putty-like consistency are composed of ore, coke, water, return-fines and flux. The bed is levelled by a cut-off plate and as the material passes under the ignition hood, hot gas is drawn downwards through the bed, eventually causing the coke particles near the surface of the bed to ignite. Once combustion is established and the ignited material emerges from under the hood, a reaction zone moves downwards through the bed under the influence of a suction fan which draws air into the surface of the bed and exhausts the outlet gas through windboxes situated below the strand. As the reaction zone works its way through the bed, the temperature of the processed material is reduced by the cool air drawn into the bed. Eventually, the reaction zone reaches the bottom of the bed. The point along the strand where this first occurs is called the burnthrough position, which is near the output end of the strand.

In all modelling exercises concerned with dynamic systems the model designer has in mind the salient features of the system which he wishes to investigate and, as a first approach, he interrelates what he considers to be the important variables and neglects those effects which appear to be of secondary importance. Considering the sinter strand as a production process, a

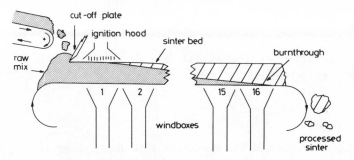

Fig. 1.1 *Schematic diagram of a sinter strand*

useful model can be formulated which relates burnthrough position to the permeability of the raw mix loaded onto the strand. In order to produce sinter of uniform quality, the variation in burnthrough position should be kept as small as possible.

1.2.1 Lumping the model and simulating the horizontal motion

The sintering process is essentially convective and since air is drawn through the bed more or less vertically, it is reasonable to lump the bed into sections along its length and assume that within a section the material is processed uniformly (Fig. 1.2). By using a large number of sections the continuous

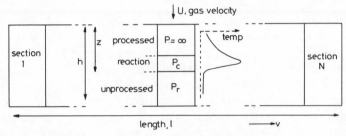

Fig. 1.2 *Sinter bed represented by lumped sections*

horizontal motion of the bed can be approximated by shifting sections incrementally.[26] For a bed composed of N sections moving at a constant horizontal velocity v the residence time of a single section in each position is θ_r where $\theta_r = l/Nv$. l is the length of the bed.

Because of the difficulty in fully instrumenting a moving bed and bearing in mind the above argument in favour of lumping, it is common practice to carry out simulated tests on static plant, usually referred to as a test pot. Effectively, the material in a test pot is equivalent to a section fixed in the bed moving at the speed of the strand. Data obtained from test pot trials are used to provide information for the simulation models.

1.2.2 Modelling a section of the bed

An empirical equation of the form

$$P = \frac{F \, h^m}{A \, s^n} \tag{1.1}$$

may be assumed for a packed bed, where
  P is the permeability of the bed
  F/A is the volumetric gas flow per unit area
  h is the thickness of the bed
  s is the suction across the bed

Voice et al.[31] showed experimentally that, for a sinter bed, it may be assumed that $m = n = 0 \cdot 6$.

Let $F/A = U$, the gas velocity. Then, from Eqn. (1.1)

$$P = U\left(\frac{h}{s}\right)^{0\cdot6} \tag{1.2}$$

or

$$s = \left(\frac{U}{P}\right)^{1\cdot67} h \tag{1.3}$$

Assume that a section of the bed comprises three zones (Fig. 1.2); unprocessed material, a reaction zone and sinter.

The permeability of the unprocessed material is assumed to remain at its initial value P_r. Sinter offers little resistance to flow and its permeability is assumed to be infinite. Within the reaction zone, the permeability P_c may be related to that of the raw mix by the equation

$$P_c = \alpha P_r + \beta \tag{1.4}$$

where the values of the constants α and β may be determined from test pot data.

Using Eqn. (1.3) the suction across the reaction zone, s_c, is given by the relationship

$$s_c = \left(\frac{U}{\alpha P_r + \beta}\right)^{1\cdot67} h_c \tag{1.5}$$

where h_c, the effective thickness of the reaction zone, is assigned a constant value determined from plant data.

At time θ, the leading edge of the reaction zone is at a depth z from the surface of the bed. The suction s_{h-z} across the unprocessed zone is related to its thickness by the equation

$$s_{h-z} = \left(\frac{U}{P_r}\right)^{1\cdot67} (h - z) \tag{1.6}$$

Therefore

$$s = s_c + s_{h-z} = U^{1\cdot67}[(\alpha P_r + \beta)^{-1\cdot67} h_c + P_r^{-1\cdot67}(h - z)] \tag{1.7}$$

Rearranging Eqn. (1.7)

$$U = \left(\frac{s}{(\alpha P_r + \beta)^{-1\cdot67} h_c + P_r^{-1\cdot67}(h - z)}\right)^{0\cdot6} \tag{1.8}$$

A fundamental assumption used in formulating the model is that the rate of advancement of the reaction zone is proportional to the gas velocity. This assumption is supported by a theoretical study carried out by Muchi and Higuchi[16] and a practical investigation reported by Boucraut and Rochas.[2]

Thus

$$T_1 \frac{dz}{d\theta} = U \tag{1.9}$$

where T_1 is a constant.

At burnthrough $(z = h)$ the permeability of the reaction zone is given by Eqn. (1.4)

$$P_c = \alpha P_r + \beta \tag{1.10}$$

From this point onwards Eq. (1.9) does not apply and the reaction zone is assumed to convert to sinter exponentially, that is

$$\frac{1}{T_2} \frac{dP}{d\theta} = P \tag{1.11}$$

with the initial condition $P = \alpha P_r + \beta$. Also Eq. (1.8) reduces to

$$U = P \left(\frac{s}{h_c} \right)^{0.6} \tag{1.12}$$

1.2.3 Fitting the equations to test pot data

A test pot consists of a box section which may have dimensions in similar proportions to those of a single pallet on the real plant (in practice, the moving grate on a sinter plant is not truly continuous but comprises a number of joined pallet sections). The conditions pertaining on-plant are simulated practically in the test pot. It is likely, however, that since the test pot operates as a batch process and there are no moving parts, there will be less leakage flow than occurs on the plant. The test pot process is monitored by flow, pressure and temperature sensors located in the windbox below the pot and by thermocouples situated at different levels inside the test pot.

By selecting from a large amount of data the results of tests in which the fuel content is the same but the permeability of the raw mix is different, a range of gas flow histories may be plotted. Values can then be assigned to the constants α, β, T_1, T_2 and h_c in the model equations to yield similar gas flow histories, Fig. 1.3. The corresponding profiles for the leading edge of the reaction zone are shown in Fig. 1.4. Normally, it would be possible to check these profiles against data obtained from the thermocouples and this aspect is considered in detail later when discussing the formulation of an analytical model. For the purposes of checking the present empirical model the data for the temperature of the exhaust gas sensed in the windbox only was used. A typical temperature record is shown in Fig. 1.5. Two features are of particular interest. Firstly, the peak temperature is very much less than the maximum temperature of sinter (360°C compared with approximately 1400°C): this is mainly due to the large heat sink formed by the grate. However, it is reasonable to assume that the time at which the peak temperature occurs is indeed related to the burnthrough time especially as the sensor is close to the grate.

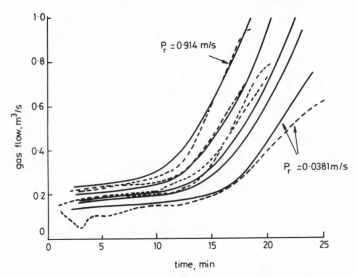

Fig. 1.3   *Gas-flow histories for test pot and model*
-------- test pot
———— model

The definition of burnthrough for the model is the point where the front of the reaction zone reaches the bottom of the bed, and as indicated in Fig. 1.2, this corresponds not to the point of peak temperature but more precisely to the point of first rise. As shown in Table 1.1, columns 3 and 4, the difference

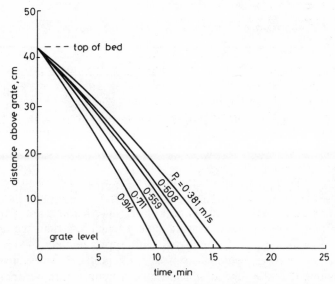

Fig. 1.4   *Reaction zone leading-edge profiles for the model*

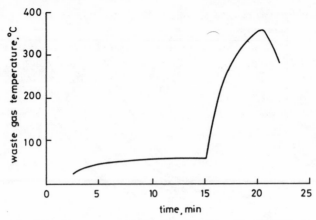

Fig. 1.5   *Temperature history of exhaust gas* $(P_r = 0.559 \; m/s)$

in time between these events is of the order of 5 minutes. A comparison of the model results (column 2) and test pot results (column 4) still shows a discrepancy of 2–3 minutes. Further inspection of the test pot data, however, (see Carter and Rose[5]) reveals an inability of the fan to maintain the desired suction at the higher flow rates encountered as burnthrough approaches. A correction factor based on the assumption that (reaction zone velocity) $\propto$ (suction)$^{0.6}$ can be applied, yielding the results shown in column 5 of Table 1.1 which are seen to agree well with those for the model (column 2). The gas flow histories shown in Fig. 1.3 also incorporate the suction correction factor.

Table 1.1   *Comparison of burnthrough times for the model and test pot.*

1	2	3	4	5
Permeability m/s	Time to burn-through MODEL (min)	Time of peak windbox TEST POT (min)	Time temperature first starts to rise TEST POT (min)	Time first starts to rise corrected for suction drop off TEST POT (min)
0·914	9·9	17·0	13·0	10·0
0·711	11·4	18·0	14·0	11·0
0·559	12·8	20·0	15·0	12·0
0·508	13·0	20·8	15·5	12·5
0·381	15·6	23·5	18·0	15·0

1.2.4 Incorporating the fan characteristics into the simulation of the moving bed

In some cases the suction across a sinter bed is maintained constant. In other cases the suction may vary with flow, typically, as shown in Fig. 1.6. The latter introduces a complication in executing a simulation run for the bed. For each time interval the sum of the flow rates through all the lumped sections of the model taken together with the value of suction for that time interval should correspond to a point on the fan characteristic. An iterative

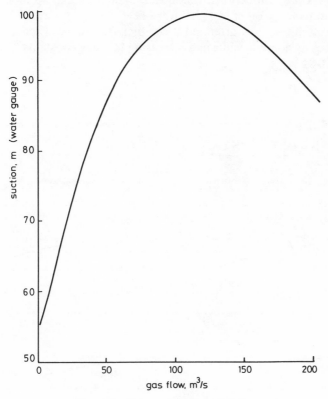

Fig. 1.6 *Exhaust fan characteristic*

procedure commencing with an estimated value of suction may be used to achieve this. Although the vanes of the fan are adjustable it is normal practice to operate at maximum setting. Consequently a single operating curve only is shown in Fig. 1.6.

1.2.5 Problems in scaling up test pot data to simulate plant

Leakage flow into the windmains on real plant is substantial and must be taken into account in comparing plant and test pot data. Carter and Rose[5] estimated a 30% leakage flow at a suction of 76 w.g. (water gauge) for one particular plant. It is probably satisfactory in calculating the leakage flow rate to assume that it is proportional to the square root of fan suction.

Experience has shown that an allowance of the above type must be supplemented by an adjustment of the model parameters in scaling up suction and bed height from test pot to plant values, otherwise the burnthrough time obtained from the simulation is likely to be different from that obtained for the plant. The reason for the discrepancy may be the variation in flow through a section of the bed as it moves from a position directly over a windbox to a position between windboxes. Similar scaling-up problems have been encountered by other workers in the field, notably Grieve.[13] Satisfactory results were achieved by Carter and Rose[5] by reducing the effective bed area a to 0·81 times its nominal value and increasing the time constants T_1 and T_2 by a factor 1/0·81.

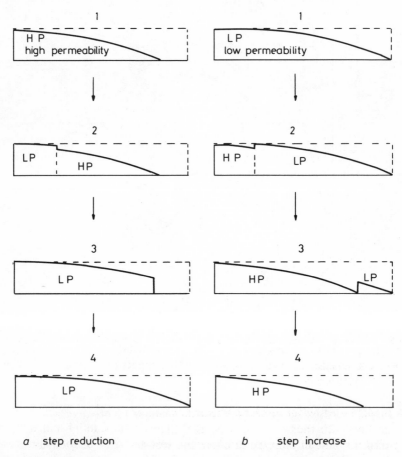

Fig. 1.7 *Progression of the leading edge of the heat wave in response to a step change in permeability*

1.2.6 Response of burnthrough position to step changes in permeability
It is logical that material of low permeability will be processed at a slower rate than material of higher permeability and vice versa. The progression of the heat-wave profiles shown sequentially in Fig. 1.7 may therefore be expected.

Figure 1.7*a* represents a step change from high to low permeability (HP to LP) which eventually yields a burnthrough position nearer to the output end of the strand. If the permeability were sufficiently low, burnthrough might, in practice, not occur at all. In terms of the model, the burnthrough time would be greater than the strand time given by l/v. The burnthrough time in this case is 'notional'. At an intermediate stage, (3), the material of high permeability has burnt through and for a short period there will be no burnthrough point, i.e. until stage (4) when the low permeability material burns through.

Figure 1.7*b* represents the reverse change from low permeability to high permeability which gives rise to relatively early burnthrough. In this case, stage (3) represents the time when two burnthrough points first occur simultaneously. This situation persists until the low permeability material is completely processed, (4). Although the phenomenon of more than one burnthrough point occurring simultaneously had previously been observed on plant it was first explained in a simulation model in 1973 by Carter and Rose.[6]

A more subtle feature of the response arising when the applied suction is governed by the fan characteristic is comparable with the non-minimum phase effect occurring in some types of control systems in that the initial movement of the burnthrough point is away from its final position and also it tends to produce a slight overshoot (Fig. 1.8).

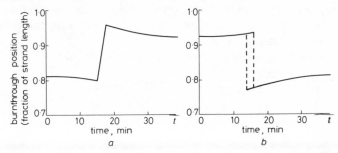

Fig. 1.8 *Response of burnthrough point to step change in raw mix permeability applied at*
t = 0
a Step reduction (0·92 – 0·76 m/s)
b Step increase (0·76 – 0·92 m/s)

1.3 A model based on fundamental chemical engineering relationships

An empirical model of the type discussed in Section 1.2 is relatively simple in concept, easy to simulate and serves as a 'black box' model in providing an appreciation of the overall behaviour of the system. Also it may be used as a basis for control system investigations concerned with the regulation of burnthrough position. The empirical model does not—and was not designed to—provide a means for the study of the internal mechanisms and reactions within the sintering process.

To gain an understanding of the processes occurring within a sinter bed and to investigate the relative importance of the process parameters to the efficient production of good quality sinter, a more detailed model based mainly on fundamental analytical relationships is required. The formulation of such a model is considered in this Section. Again, it is convenient to consider a test pot as representing a section of the bed.

The input information should correspond more directly to that which is considered important in test pot investigations. For example, for the raw mix:

    ore content   (%)
    coke content   (%)
    water content   (%)
    limestone content (%)
    return-fines content (%)
    ore characteristics
    size analysis of mix
    bulk density of mix

and for the test pot:

    depth of bed
    fan suction
    temperature of ignition gases
    ignition period.

Permeability, which for the empirical model was the only input variable used to represent the raw-mix characteristics, may be regarded as an intermediate measured variable on plant and is not required for this model.

Output measurements on the test pot include:

    temperature at different levels and in the windbox
    exit gas velocity
    concentration of carbon dioxide in the exit gas (C_{CO_2}).

Sinter quality tests carried out in practice include:

    shatter
    abrasion
    reducibility

and

    low temperature breakdown.

Unfortunately a mathematical model cannot yield corresponding information but it is postulated that sinter quality is related to the amount of fusion occurring in the process and that this can be assessed from the temperature response of the model. Young,[34] in considering the performance of a model similar to the one discussed in this section, includes tentative results relating a strength index to sintering conditions.

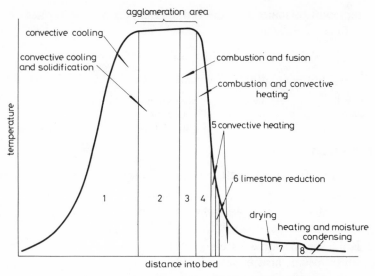

Fig. 1.9 *Temperature profile showing process phases*

A zonal structure showing the main phases of the process may be assumed (Fig. 1.9). The zones, however, do not have well-defined boundaries but rather develop as the simultaneous solution of the process equations progresses, allowing combustion to die out naturally as all the carbon particles react.

1.3.1 Mass and energy balance equations
The overall combustion of coke particles is dependent upon both chemical and mass transfer because, before combustion of coke can occur, oxygen has to diffuse to the surface of the coke particle. The overall combustion rate R_c is given by the equation

$$R_c = 4\pi r_c^2 n_c K C_{O_2} \tag{1.13}$$

where $4\pi r_c^2$ is the surface area of the coke particles
  n_c is the number of coke particles per unit volume
  $K = K_c K_m / (K_c + K_m)$
  K_m is the coefficient of mass transfer
  K_c is the chemical combustion rate constant

Using basic chemical engineering principles, equations for the heating, cooling and combustion zones may be derived.[11, 26]
The mass velocity of the gas through the bed is given by

$$G = \epsilon \rho_g (\partial z / \partial \theta) \tag{1.14}$$

where ϵ represents bed voidage ($=$ volume of voids/specific volume of material)
  ρ_g represents density of the gas

z represents distance into the bed, measured from the surface

θ is time

The energy balance equation for the gas is

$$G \cdot \partial(c_g\tau)/\partial z + \epsilon\rho_g \, \partial(c_g\tau)/\partial\theta + h_p a(\tau - t) = 0 \qquad (1.15)$$

| bulk convection | accumulation in gas | convection exchange to solid |

and for the 'solid'

$$(1 - \epsilon)\rho_p \, \partial(c_p t)/\partial\theta + h_p a(t - \tau) - H_c \cdot R_c = 0 \qquad (1.16)$$

| accumulation in solid | convection exchange to gas | heat from reactions in solid |

where a is the specific surface area

c_g is the specific heat of gas

c_p is the specific heat of solid

H_c is the heat of combustion in coke

h_p is the convective heat transfer coefficient between solid particles and gas

ρ_p is the density of the solid

t is the temperature of solid

τ is the temperature of gas

Considering the mass balance of oxygen and carbon dioxide in the gas phase

$$\partial(G \cdot C_{O_2}/\rho_g)/\partial z + \partial(\epsilon C_{O_2})/\partial\theta + R_c = 0 \qquad (1.17)$$

| bulk convection | accumu- lation in gas | net change in C_{O_2} by reaction |

$$\partial(G \cdot C_{CO_2}/\rho_g)/\partial z + \partial(\epsilon C_{CO_2})/\partial\theta - R_c = 0 \qquad (1.18)$$

| bulk convection | accumulation in gas | net change in C_{CO_2} by reaction |

and the following formulae may be used to calculate the gas concentrations and densities

$$C_{O_2} + C_{CO_2} + C_{N_2} = 44 \cdot 6(273/\tau) \qquad (1.19)$$

$$\rho_g = (32C_{O_2}^* + 44C_{CO_2}^* + 28C_{N_2}^*)10^{-3}(273/\tau) \qquad (1.20)$$

where C_{O_2}, C_{CO_2}, C_{N_2} are the concentrations of O_2, CO_2 and N_2 in the gas, and C^* indicates concentration at N.T.P.

Assumptions made in formulating Eqns. (1.13) to (1.20) are:

1. Gas flow is one-dimensional.
2. Gas species are ideal.
3. Eddy diffusion effects are negligible.
4. The system is adiabatic.
5. Radiative effects within the gas and between the gas and solid are insignificant.
6. The reaction of coke with oxygen takes place on the surface of the coke particles and not in the gas.
7. The temperature inside a solid particle is identical to the temperature at the surface. Saunders and Ford[27] have shown experimentally that conduction within small particles (approximately 6 mm diameter) in a packed bed is sufficiently large to allow temperature gradient effects to be neglected.
8. Convective heat transfer is the major factor promoting heat-wave propagation. Axial heat conduction and radiation are assumed negligible.

1.3.2 Drying and condensation
In formulating the model for any part of a process the modeller inevitably has in mind the following questions:

(*a*) how simple can he make the model without introducing significant errors in the behaviour of the process?
(*b*) if a complex formulation were used, would adequate pilot plant data be available to establish the equation coefficients? and
(*c*) how practicable would the simulation of a complex model be?

With respect to the drying phase of the sintering process it is possible to formulate energy and mass balance equations, but the rate constant for drying at the front edge of the heat wave is of the same order of magnitude as the update period of the digital simulation which is perfectly acceptable for the other phases of the process. It is unlikely, at least from normal tests, that data for the proportions of water extracted during the stages of drying (of which there are three[7]) will be available. Also, the removal of water during sintering accounts for approximately 10% only of the available energy.

All these points lead to the use of a simplified approach in dealing with the drying phase. It is assumed that all evaporation occurs at 100°C and that

Heat from gas = (mass of water evaporated)

× (latent heat of vaporization)

As the gas passes down the bed laden with moisture from the drying zone it cools until it reaches the dew-point. At the dew-point, the moisture condenses and the bed temperature rises. Using this approach, the raw mix below the combustion and heating zones gradually rises in temperature until it is all at the dew-point temperature. Using a least-squares curve-fitting technique and graphical data given by Perry and Chilton[21] the following relationship between dew-point temperature τ_{dew} and moisture content may be derived

$$\tau_{dew} = 293\!\cdot\!4 + 324\!\cdot\!6\ W - 594\!\cdot\!1\ W^2 + 292\!\cdot\!1\ W^3 \tag{1.21}$$

where W is the humidity of the gas.

The heat gained by the bed due to condensation is given by the equation

Heat to bed = (mass of water condensed)

× (latent heat of condensation)

Drying occurs at a temperature between the dew-point and the boiling-point temperatures. In a sinter bed the dew-point is at 60°C, so drying occurs between 60°C and 100°C. The assumption made in the model that all drying occurs at 100°C is therefore slightly in error. (It has been calculated that if all drying occurred at the other extreme of 60°C this would give rise to a 1% discrepancy only in energy balance.)

1.3.3 Fusion

The heat produced from the combustion of the coke causes the temperature to rise to a value in the range 1200–1400°C and bonding takes place between the particles. Work of a practical nature and particularly that reported by Nyquist[17] and Price and Wasse[23] illustrates the difficulty of predicting mineralogy changes due to fusion. Analytical modelling of the fusion phase is extremely complex. It is therefore at this stage of the modelling exercise that one seeks an alternative approach. At least two authors of learned papers in the past have assumed that the heat produced in combustion causes the bed temperature to soar far beyond that which is achievable in practice. The approach used here, attributable to Dash and Rose,[8] is to allow the temperature to rise to a prescribed value (in the range 1200–1400°C) and to store the excess heat, assuming it to be used in fusion, until the cool air drawn through the bed behind the combustion zone causes the temperature to drop. The validity of the model depends substantially upon this assumption which is founded on the fact that the maximum temperature achieved in practice— at least for high-quality self-fluxing mixes containing hematite and magnetite—is reasonably independent of the coke content. The majority of sintering mixes are of this type.

1.3.4 Limestone reduction

The reduction of limestone is an endothermic reaction and occurs at temperatures in excess of 600°C. It is estimated that about 5% of the available energy is used in this part of the process.

1.3.5 Formulae used to determine coefficients in the foregoing equations

1.3.5.1 *Heat and mass transfer.* The coefficient of convective heat transfer h_p is dependent upon the physical properties of the solid and gas. Many formulations of the heat transfer coefficient are based on the following relationship between the Nusselt, Reynolds and Prandtl numbers N_u, R_e and P_l respectively

$$N_u \epsilon \propto P_l^n R_e^m$$

where $N_u = h_p d_p / \lambda$, $R_e = d_p G / \mu$ and $P_l = c_g / \lambda$
d_p is the effective pellet diameter (see Section 1.3.9)
λ is the thermal conductivity of the gas
μ is the viscosity of the gas

For a sinter bed it is considered[16] that the most appropriate equation is

$$N_u \epsilon = 2 \cdot 0 + 0 \cdot 71 \, P_l^{1/3} R_e^{1/2} \tag{1.22}$$

A similar formula exists for the coefficient of mass transfer between a gas and particles in a packed bed, namely

$$S_h \epsilon = 2 \cdot 0 + 0 \cdot 71 S_c^{1/3} R_e^{1/2} \tag{1.23}$$

where $S_h = K_m d_p / D$, $S_c = \mu / \rho_g D$
D is the diffusion coefficient.

1.3.5.2 Specific heat. The following formula may be assumed for the specific heat of sintering gases,[24]

$$c_g = A + B\tau - C\tau^2 \tag{1.24}$$

where, for a mixture composed of N_2, O_2 and CO_2 in the ratio $0 \cdot 8 : 0 \cdot 1 : 0 \cdot 1$,

$$A = 880 \cdot 0 \qquad B = 0 \cdot 31 \qquad C = 8 \cdot 0 \times 10^{-5}$$

Changes in the $O_2 : CO_2$ ratio have negligible effect on the specific heat.
   The specific heat of solid material of the type used in sintering[21] is given by the equation

$$c_p = D + Et \tag{1.25}$$

where $D = 753 \cdot 0$ and $E = 0 \cdot 25$.

1.3.5.3 Gas viscosity. The following equation (Sutherland's formula) may be used to estimate the viscosity of gas at different temperatures:

$$\mu = \mu_0 \left(\frac{\tau}{273} \right)^{3/2} \frac{C + 273}{C + \tau} \tag{1.26}$$

where μ_0 is the viscosity at 0°C and C is a constant, the value of which depends on the gas type. For air, $C = 113$ and $\mu_0 = 1 \cdot 72 \times 10^{-5}$ kg/ms.

1.3.5.4 Diffusion coefficient of O_2 through N_2. The value of the diffusion coefficient of oxygen through nitrogen D_{ON} is given by Parker and Hottel[19] and others as

$$D_{ON} = 1.8 \times 10^{-5} \left(\frac{\tau}{273}\right)^{1.5} \tag{1.27}$$

1.3.5.5 Coke combustion. Parker and Hottel[19] give the following formula for the chemical combustion rate K_c:

$$K_c = 6.52 \times 10^5 \, e^{-18,500/RT} \sqrt{t} \tag{1.28}$$

From Eq. (1.23)

$$K_m = \frac{D_{ON}}{\epsilon d_p} [2.0 + 0.71 \, S_c^{1/3} R_e^{1/2}]$$

This equation is based on the assumption that coke is converted completely to carbon dioxide. Schluter and Bistranes[28] introduces a factor ϕ, the value of which depends on the proportion of carbon monoxide contained in the exhaust gas. Hence

$$K_m = \phi \frac{D_{ON}}{\epsilon d_p} [2.0 + 0.71 \, S_c^{1/3} R_e^{1/2}] \tag{1.29}$$

where, for a $CO_2:CO$ ratio of $4:1$, $\phi = 1.11$.

1.3.6 Method of simulation for gas-solid heat exchange

The simulation method is best described initially for the case of pure convective heat transfer between a cold bed and hot gas, neglecting the phases representing drying, combustion, fusion and limestone reduction.

The bed is assumed to be divided vertically into n zones (Fig. 1.10). The thickness of the ith zone is Δz_i and therefore $\sum_{i=1}^{n} \Delta z_i = h$ where h is the thickness of the bed. Consider a narrow band of hot gas of thickness δz, $(\delta z \ll z_i)$ passing downwards through each zone of the bed in turn. Because of the large difference in density between the solid and the gas (approximately $3000:1$) it is admissible to allow the temperatures $t_1 \cdots t_n$ of the solid material in the zones 1 to n respectively to remain constant for a prescribed period $\Delta\phi$ (which greatly exceeds the time taken for the band of gas to make one pass through the bed). Calculations are carried out for a large number of passes, allowing temperature profile to virtually reach a steady state before the solid temperatures $t_1 \cdots t_n$ are updated (after each $\Delta\phi$ interval). From Eqn. (1.14), the time taken for the band of gas to pass through the ith zone is $\Delta\theta_i$, where

$$\Delta\theta_i = \frac{\epsilon \rho_g \, \Delta z_i}{G} \tag{1.30}$$

Neglecting the gradient across the band of gas, Eqn. (1.15) reduces to the form

$$\epsilon \rho_g \, d(c_g \tau)/d\theta + h_p a(\tau - t) = 0 \tag{1.31}$$

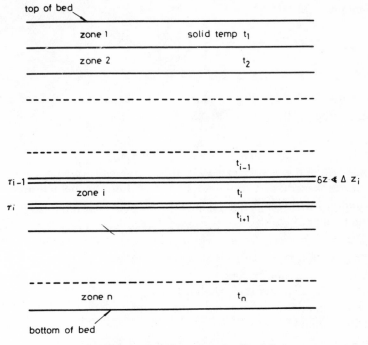

Fig. 1.10   *Zonal structure for simulation*

Therefore

$$\epsilon \rho_g \frac{d(c_g \tau)}{d\tau} \frac{d\tau}{d\theta} = h_p a(t - \tau) \tag{1.32}$$

Using the expression for c_g given in Eqn. (1.24), the temperature τ_i of the gas leaving the ith zone and entering the $(i + 1)$th zone is therefore given by the equation

$$\int_{\tau_{i-1}}^{\tau_i} \frac{(A + 2B\tau - 3C\tau^2)\epsilon \rho_g}{h_p a(t_i - \tau)} \, d\tau = \int_0^{\Delta \theta_i} d\theta \tag{1.33}$$

Given the initial temperature τ_0 of the gas entering the bed and the initial temperatures of the solid $t_1 \cdots t_n$, the temperature history of a band of gas passing through the bed may be calculated. After a large number of passes spanning the period $\Delta \phi$ the solid temperatures $t_1 \cdots t_n$ are then updated using the relevant terms in Eqns. (1.15) and (1.16), namely

$$G \frac{d(c_g \tau)}{d\tau} \frac{d\tau}{dz} = h_p a(t - \tau) \tag{1.34}$$

and

$$(1 - \epsilon)\rho_p \frac{d(c_p t)}{dt} \frac{dt}{d\theta} + h_p a(t - \tau) = 0 \tag{1.35}$$

Incrementally

$$t_{i_{\theta+\Delta\phi}} = t_{i_\theta} + \frac{Gc'_g(\tau_{i-1} - \tau_i)\,\Delta\phi}{(1 - \epsilon)\rho_p c'_p\,\Delta z_i} \tag{1.36}$$

where $c'_g = A + 2B\tau - 3C\tau^2$ and $c'_p = D + 2Et$

Repeated use of the above procedure yields the gas and solid temperature histories for the heat-exchange process.

1.3.7 Extension of the simulation method to sintering

The modelling of the various phases of the sintering process has already been discussed. Implementation of the model is based on the method used for the simulation of the heat exchanger where the ith zone may lie in any one of the following regions (Fig. 1.9):

1. cooling
2. cooling and solidification
3. coke combustion and fusion
4. convective heating and coke combustion
5. convective heating
6. limestone reduction
7. drying
8. heating and condensing

The temperature of the gas in all regions is given by Eqn. (1.33) with t_i assigned appropriate values.

For regions (1) and (5), where straightforward convective cooling and heating occur, Eqn. (1.36) may be used to update the solid temperature.

In regions (2) and (3) the solid temperature is held at the constant value t_F. The heat available for fusion F_u in region (2) is given by

$$F_{u_{i,\theta+\Delta\phi}} = F_{u_{i,\theta}} + Gc'_g(\tau_{i-1} - \tau_i)\,\Delta\phi \tag{1.37}$$

and in region (3) by

$$F_{u_{i,\theta+\Delta\phi}} = F_{u_{i,\theta}} + Gc'_g(\tau_{i-1} - \tau_i)\,\Delta\phi + \Delta z_i R_{c_i} H_c\,\Delta\phi \tag{1.38}$$

The solid temperature update equation for region (4) is

$$t_{i,\,\theta+\Delta\phi} = t_{i,\,\theta} + \frac{Gc'_g(\tau_{i-1} - \tau_i)\,\Delta\phi}{c'_p(1 - \epsilon)\rho_p\,\Delta z_i} + \frac{R_{c_i} H_c\,\Delta\phi}{c'_p(1 - \epsilon)\rho_p} \tag{1.39}$$

For region (6), the total heat needed for limestone reduction L_R is given by the product of the mass of limestone in a zone and the heat needed to reduce a unit mass of limestone. The following equation may be used to simulate limestone reduction

$$L_{R_{i,\theta+\Delta\phi}} = L_{R_{i,\theta}} - Gc'_g(\tau_{i-1} - \tau_i)\,\Delta\phi \tag{1.40}$$

For the drying region (7)

$$X_{w_{i,\theta+\Delta\phi}} = X_{w_{i,\theta}} - Gc'_g(\tau_{i-1} - \tau_i) \, \Delta\phi/H_v \tag{1.41}$$

where X_w is the mass of water and H_v is the latent heat of vaporization.

For region (8), heating and condensing, the solid update temperature is given by the equation

$$t_{i,\theta+\Delta\phi} = t_{i,\theta} + \frac{Gc'_g(\tau_{i-1} - \tau_i)}{c'_p(1 - \epsilon)\rho_p \, \Delta z_i} + \frac{W_{G_{i,\theta}} H_v \, \Delta\phi}{c'_p(1 - \epsilon)\rho_p \, \Delta\theta_i \, \Delta z_i} \tag{1.42}$$

When the bed reaches the dew-point no heat is supplied by the condensation of water vapour. $W_{G_{i,\theta}}$ is the mass of water condensed from the gas during the gas transit time $\Delta\theta_i$.

The overall combustion rate of coke R_c is dependent on the concentration of oxygen in the gas surrounding the burning coke particles. As the sinter gas passes through the combustion regions, oxygen is consumed, with the result that the combustion rate of coke in the lower part of the combustion regions is decreased through a lack of oxygen. Therefore, for the processes in regions 1 to 8, changes in oxygen concentration should be represented. The mass balance of the oxygen concentration at any point is given by Eqn. (1.17). If changes in the bed voidage due to combustion are ignored, the concentration of the gas leaving the ith zone and entering the $(i + 1)$th zone is given by

$$C_{O2,i} = C_{O2,i-1} - R_{c_i} \, \Delta\theta_i/\epsilon \tag{1.43}$$

The gas concentration at N.T.P. is given by the equation

$$C^*_{O2,i} = C^*_{O2,i-1} - R_{c_i} \, \Delta\theta_i\tau_i/273\epsilon \tag{1.44}$$

1.3.8 Some results of simulation exercises

Simulation investigations in which a history of gas flow is assumed are useful in developing the model and in assessing the relative effects of changing the values of some of the process parameters. With parameters set at values estimated from test pot trials the heat-wave leading edge profiles shown in Fig. 1.11 were obtained and compared with measurements from thermocouples situated at different depths in the bed. The case investigated is for a hematite mix with 4·5% coke content. At this stage, gas flow takes the form of an input to the system and the values used were those recorded in the test pot trials. The heat-wave profiles from the simulation correspond closely to the measured values and provide the required confidence in the model. The form in which the profiles are presented tends to emphasise quite small discrepancies in the position at which the temperature is measured. A discrepancy of 0·01 m between the nominal depth and the true depth of the sensing element inside the material could bring the two sets of curves much closer together. Bearing in mind the finite diameter of the thermocouple probes (0·006 m) and

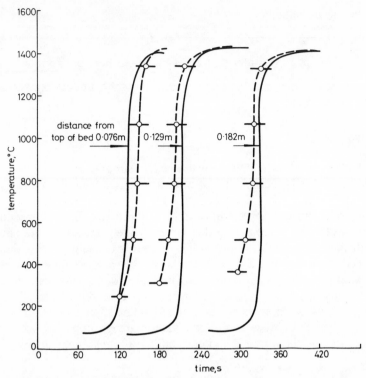

Fig. 1.11   *Heat wave leading-edge profiles; comparison between test pot thermocouple measurements and simulation results*
        -------- test pot
        ———— simulation

the forces on the probes due to the reacting material, a 1 cm discrepancy is quite feasible. Also, the test pot material may not have been entirely homogeneous.

The model provides a useful means for investigating the effects of manipulating individual process variables with others held constant. The manipulation of variables such as airflow and voidage have two basic effects; one is to change the heat-wave propagation velocity, the other is to change the width of the fusion zone. The velocity of the heat wave through a sinter bed should be as high as possible if a sinter plant is to produce a maximum output; however, if the fusion zone does not produce the right degree of agglomeration there may be a large quantity of rejected sinter and hence a reduced productivity. Too wide a fusion zone will cause large lumps of sinter to form, which will tend to block the passage of air through the bed and hinder the advancement of the heat wave. Too narrow a fusion zone results in a weak fretted structure and a high return fines ratio. The required agglomeration conditions may vary from ore type to ore type and also depend upon the lime/silica ratio; however the following results give an insight into the effect of

Table 1.2 *Input data for simulation investigations*

	Fig. 1.11	Fig. 1.12	Fig. 1.13		Fig. 1.14			
Ignition temperature	977–1377	1180	1180	1180	1180	1180	1180	°C
Ignition time	40	25–50	40	40	40	40	40	s
Voidage	0·4	0·4	0·4	0·25–0·50	0·4	0·4	0·4	—
Gas inlet velocity	1	1	1	1	0·8–2·0	1	1	m/s
Initial bed temperature	0	0	0	0	0	0	0	°C
Δz, width of zones	0·0025	0·0025	0·0025	0·0025	0·0025	0·0025	0·0025	m
$\Delta\Phi$, update period	1	1	1	1	1	1	1	s
Density (return fines and ore)	5000	5000	3500–6000	5000	5000	5000	5000	kg/m³
Density of coke	1000	1000	1000	1000	1000	1000	1000	kg/m³
% coke	5	5	5	5	5	3–5·5	5	—
% water	5	5	5	5	5	5	0–18	—
Diameter of particles	0·0030	0·0030	0·0030	0·0030	0·0030	0·0030	0·0030	m

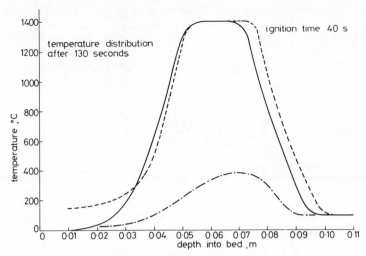

Fig. 1.12   *Effect of changing ignition gas temperature*
Ignition temperatures
-------- 1380°C
———— 1030°C
—·—·— 980°C

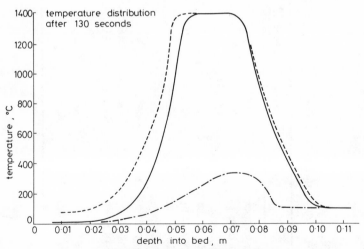

Fig. 1.13   *Effect of changing the 'ignition period'*
Ignition time
-------- 40 s
———— 30 s
—·—·— 25 s

manipulating individual process variables. The input data used for calculating the results for a typical 0·25 m bed are given in Table 1.2.

1.3.8.1 Effect of changing ignition gas temperature. Figure 1.12 shows the temperature distribution within the bed after 130 seconds, for three different ignition temperatures.

The main points arising from these results are:

(i) Propagation of the heat wave is only slightly affected by the ignition temperature.
(ii) The bed would not ignite at 980°C whilst at temperatures above 1030°C sintering was successfully commenced.

1.3.8.2 Effect of changing the ignition period. The effect of longer ignition periods is to slightly widen the fusion zone (Fig. 1.13). Also the point at which ignition commences is dependent on the ignition time.

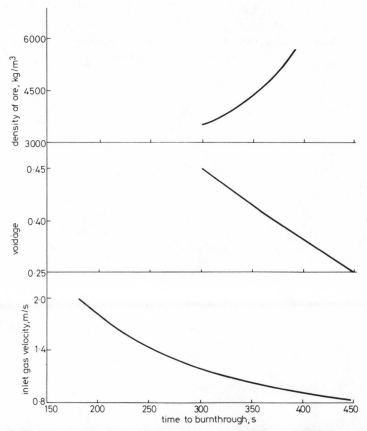

Fig. 1.14 *Effect of density, voidage and gas velocity changes on burnthrough time*

1.3.8.3 Effects of changing the density of the mix, voidage and gas velocity. Figure 1.14 shows the effect of changes in voidage, gas velocity and density of the mix on burnthrough time. Changes in gas velocity have a very marked effect on heat-wave propagation velocity, whilst voidage and density changes make less, though still significant, changes to the burnthrough time. Changes to these variables appear to have little effect on the width of the fusion zone.

1.3.8.4 Effect of varying the coke and water content of the raw mix. Variations in coke and water content whilst not altering the burnthrough time have a marked effect on the width of the fusion zone (Fig. 1.15). When simulating low contents (less than 3·5%) a longer ignition period proved necessary.

1.3.8.5 Influence of particle diameter. Variations in the effective diameter of the raw-mix particles do not affect either the width of the combustion zone or the burnthrough time. However, if the mix is larger than 5 mm, ignition will not occur unless the ignition time is increased.

1.3.9 Inclusion of gas flow calculations

Care must be taken in interpreting simulation results from a model in which gas flow is used as an input and a further stage of modelling is necessary to

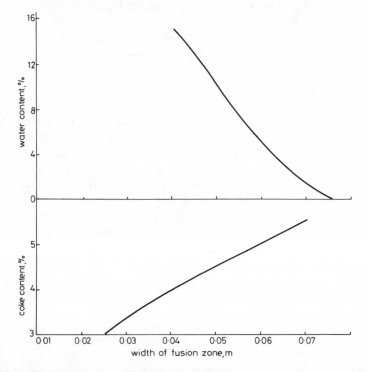

Fig. 1.15 *Effect of changes in coke and water content of the raw mix on the width of the fusion zone*

relate gas flow to the suction across the bed. Empirical relationships between flow and suction were discussed in Section 1.2.2. Several investigators have derived analytical equations relating the flow of gas through a packed bed to the pressure drop across the bed. Carman[4] and Ergun[12] in particular have made contributions in this field. Perhaps the most appropriate equation for use here is that developed by Szekely and Carr[29] for non-isothermal flow, namely

$$\frac{G^2}{\epsilon^2} \ln \frac{V}{V_i} + \int_i (AG\mu + BG^2)\, dz = -\int_i \frac{dP}{V} \tag{1.45}$$

where $V = 1/\rho_g$ (the specific volume)

$$A = \frac{150(1 - \epsilon)^2}{d_p^2 \epsilon^3}$$

$$B = \frac{1 \cdot 75(1 - \epsilon)}{d_p \epsilon^2}$$

Provided that the properties of the solid (effective particle diameter and void-age) and of the gas (density, viscosity and temperature) from inlet to exit are known, Eqn. (1.45) may be used to determine the gas-flow history from ignition to burnthrough.

For isothermal flow of an incompressible fluid Eqn. (1.45) reverts to a simpler form usually known as Ergun's equation

$$\frac{\Delta P}{h} = A\mu U + B\rho_g U^2 \tag{1.46}$$

Equation (1.46) may be applied to a sinter bed (or test pot) before ignition, because here changes in gas volume caused by the pressure gradient across the mix at sintering suction have a negligible effect on the gas flow (1–2%).

A measurement of the bulk density and real density of the mix at the loading stage gives an estimate of the initial voidage. When suction is applied the bed compacts, causing changes in the voidage and a reduction in bed permeability. This is why, in practice, it proves uneconomic to sinter at a suction head greater than 1·5 m water. The effective pellet diameter of the input mix may be calculated from Ergun's equation by measuring the gas flow at a low suction and assuming that the voidage remains unchanged. Provided that the input mix is satisfactorily balled the value of d_p does not change with increasing suction, therefore a second measurement of the gas flow at sintering suction gives the compacted voidage of the raw mix.

Effective pellet diameter of the input mix is related to actual pellet size by the formula

$$\frac{1}{d_p} = \frac{1}{\alpha'}\left(\frac{f_1}{d_1} + \frac{f_2}{d_2} + \cdots + \frac{f_n}{d_n}\right) \tag{1.47}$$

where f_n is the fraction of pellets between two sieve sizes with a mean size d_n and α' is a shape factor (since the pellets are not necessarily spherical).

Measurements of α', d_p and ϵ for different mixes of the same ore, containing different amounts of return-fines and coke, have shown that the values of α' and ϵ are nearly constant. It may be deduced, therefore, that changes in initial bed permeability are due to changes in the size of the input mix.

Because the simulation method is based on dividing the bed into n layers Eqn. (1.45) becomes

$$G^2 \sum_1^n \frac{V}{\epsilon^2} \ln \frac{V_0}{V_i} + G \sum_1^n VA\mu \, \Delta z + G^2 \sum_1^n VB \, \Delta z = -\sum_1^n dP \qquad (1.48)$$

Also

$$\sum_1^n \Delta z = h \quad \text{and} \quad -\sum_1^n dP = \Delta P$$

The gas temperatures, carbon dioxide concentrations and gas humidities given by the simulation of the energy and mass balance equations are used to calculate the gas density and viscosity in each zone.

Voidage is assumed to remain constant during sintering, i.e. any voids formed by volatile action (in drying, limestone reduction and coke combustion) are trapped within the cooling sinter. Particle growth during sintering is

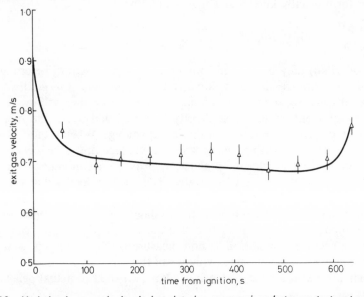

Fig. 1.16 *Variation in gas velocity during sintering; comparison between test-pot measurements and simulation results*

  ⌇   test pot
———   simulation

simulated through the equation

$$(d_p)_F = (d_p)_I(1 + bF_{u, max}) \tag{1.49}$$

where $(d_p)_I$ is the initial value of effective particle diameter
 $(d_p)_F$ is the final value of effective particle diameter
 $F_{u, max}$ is the maximum heat available for fusion
 b is the growth constant

Simulation results using the above method of calculating gas flow are compared with test pot data in Fig. 1.16. Bearing in mind that measured exit velocities are subject to an error of 0·01 to 0·02 m/s the results show good agreement.

1.3.10 Factors affecting gas flow through a sinter bed
The model incorporating Szekely and Carr's equation provides a useful method of investigating the factors which affect the rate of gas flow through the bed. The importance of building into the simulation correctly changing gas density, gas viscosity and agglomeration of the mix is illustrated by comparing the gas-flow curve of Fig. 1.16 with the results obtained from simulation exercises in which each of these variables is held constant (Fig. 1.17). If d_p remained constant during sintering, gas flow would be restricted causing a long burnthrough time. If gas density and viscosity were each held constant and d_p were allowed to change, larger gas flows would occur. The results indicate that in the true model, the tendency for changes in density and viscosity to restrict the gas flow is countered by the increase in particle size during fusion.

   When suction is applied across the raw mix the pressure gradient across the bed is constant, but once the bed is ignited the gradient at any point changes

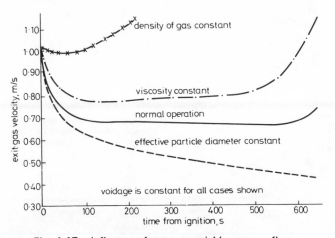

Fig. 1.17 *Influence of process variables on gas flow*

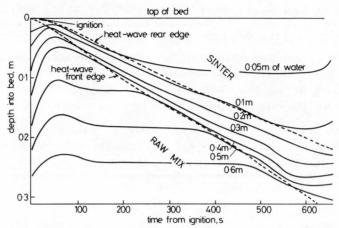

Fig. 1.18 *Pressure contours for a sinter test pot from ignition to burnthrough*

with time (Fig. 1.18). The position of the heat wave is indicated in the diagram. (The heat wave is defined here as that part of the bed which is over 1000 K.) As the heat wave passes through the bed, the pressure dropped across the heat wave increases and the pressure dropped across the raw mix decreases. At burnthrough nearly all the pressure is dropped across the heat wave. The sintered material behind the heat wave offers less resistance to flow than the raw mix and heat-wave zones of the bed.

For the ore studied, the voidage was assumed to remain constant during sintering. For ores having a high volatile content, e.g. English ores, both the effective particle diameter and the voidage increase during sintering. Figure 1.19 shows how the gas flow would change if the increased voidage

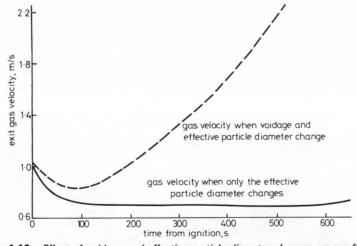

Fig. 1.19 *Effect of voidage and effective particle diameter changes on gas flow*

caused by drying and combustion had resulted in free (as opposed to trapped) voids in the agglomerate. The gas flow increases if both the voidage and effective particle diameter change during sintering.

This indicates that the way in which sinter is formed affects the rate of gas flow through the bed and hence the rate of sinter production.

Finally, it should be noted that although the more complex form of Eqn. (1.45) is used for the calculation of gas flow, the term $G^2/\epsilon^2 \ln V/V_i$ contributes less than 3% to the value of gas flow obtained and the simpler version known as Ergun's equation for isothermal flow could be used throughout without producing a significant error. This deduction is in line with recent studies on the flow through porous media, particularly by Koh et al.[15]

1.4 Sinter strand control

During the past 15 years particular attention has been paid to the problems of control and automation of the sintering process especially where large new sinter plants have been installed. An interesting evaluation of control methods used in a number of European works was published by Bragard[3] in 1967.

It can be generally stated that the purpose of using stabilization and automatic control devices is to produce a sinter of constant quality under optimum conditions. Although the blast furnace operator may give more importance to one quality compared with others, e.g. strength, reducibility, basicity, chemical composition, his general demand is for a sinter product with constant properties allowing smooth blast furnace operation.

Poos et al.[22] present data showing the dependence of raw mix permeability on moisture content and sinter quality on the proportion of coke in the raw mix. Figure 1.20 shows a plant layout which uses proprietary instruments to sense permeability and sinter quality in closed-loop control. In this approach, which has become fairly standard in the industry, the emphasis is on correcting the mix to achieve the required output.

The regularity of sinter chemistry has also been improved during recent years by adopting better blending techniques with the replacement of old disc feeders by belt weighers and the semi-continuous control of chemical composition of the raw mix by X-ray equipment.

Despite the use of these methods the effective pellet diameter (or permeability) of the raw mix and, hence, burnthrough position do vary. The author and his collaborators have, during the past few years, focused their attention on the problems of controlling burnthrough position by manipulating plant variables. It was considered, in the first instance, that there were three possible parameters available, namely suction applied across the bed, the height of the bed and strand speed, and two control philosophies, feedback of the measured burnthrough position and feedforward using measured raw mix

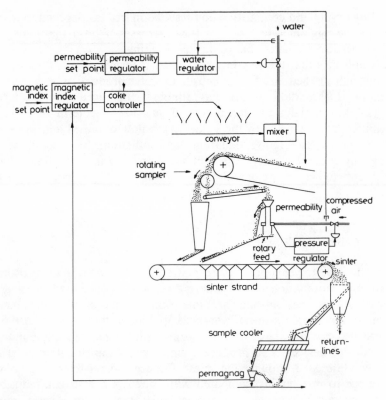

Fig. 1.20   *Control systems commonly used in present-day sinter plants*

permeability (Fig. 1.21). Simulation studies[25] in which an empirical model of the process was assumed, led to the conclusion that best results could be achieved by feedforward control operating on the cut-off plate to adjust bed height. Bragard[3] reported that bed height had been used as an action variable in processing Lorraine ores which give a highly permeable bed and allow a wide range of adjustment to bed height. On the other hand, Pengelly and Carter[20] comment that the 'control of feed packing or of bed height is of questionable value in that the variations in feed rate or strand speed tend to defeat the objective of consistent operation and, therefore, a consistent product'.

The simulation studies showed that feedback or feedforward control through fan suction or strand speed were theoretically of equal merit. The manipulation of suction through the adjustment of the fan vanes or fan speed was largely ruled out because the engineering problems could be difficult and, in any case, if there were an allowance made for increasing fan suction for control purposes it could be argued that by running the fan at full capacity, a corresponding increase in feed rate and strand speed could be made to in-

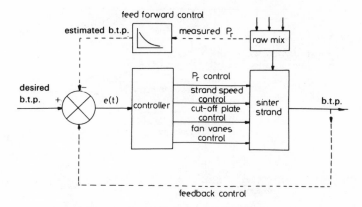

Fig. 1.21 *Possible control strategies*

crease the production rate of the plant. Control through strand speed has recently been investigated in some detail (Dash and Rose[9]) and it is this aspect which is dealt with in the remaining sections of this chapter. Bragard[3] refers to the need to adjust strand speed and feed rate to regulate burnth-rough position but it was considered that through incorporating permeability control, adjustments to strand speed would be infrequent. Kinsey,[14] in discus-sing the design features of a modern sinter plant, includes a description of a speed control loop in which strand speed is varied automatically by a burn-through control circuit and the feed rate of material onto the strand is adjusted by transmitting to the weigh belts under the raw material bins a signal based on the product of strand speed and bed height (indicating the volume being processed on the strand).

Feedback control based on the measurement of burnthrough position would be attractive if the latter could be measured accurately. In practice, a rather crude estimate of burnthrough position is obtainable from measure-ments of the temperature of the waste gases in the windboxes at the output end of the strand (Bragard[3]).

1.5 Burnthrough control by manipulating strand speed; a simulation investigation

A feedforward control scheme is considered here in which it is assumed that a perfect model of the process is available and is used to continually predict burnthrough from measurements of the effective mean pellet diameter of the raw mix. However, as is later explained, the system corresponds precisely to a feedback control scheme in which true measurements of the height of the flame front at positions along the strand are available.

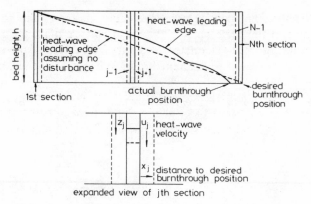

Fig. 1.22   *Schematic diagram of sinter bed represented by N sections*

1.5.1 Modelling for the control investigations

A very much simplified model was used for the control investigations, (*a*) because a strong correlation between particle diameter and burnthrough position has been established from plant data and (*b*) use of a more complex model of the process would consume an intolerable amount of computer time.

The analytical model was used to derive a relationship between the effective mean diameter of the pellets and the burnthrough time θ_b (for a strand operating with constant bed height and fan suction). For the range of mean pellet diameters that occurred during strand operation

$$\theta_b = A - B \, d_p \tag{1.50}$$

where A, B are constants ($A = 1650$, $B = 410$).

The strand is assumed to be divided into N sections, each moving horizontally along the strand at a controllable speed (Fig. 1.22). Considering the jth section, the velocity of the heat wave u_j within this section is assumed to be constant. Then

$$u_j = \frac{h}{A - Bd_{p_j}} \tag{1.51}$$

If z_j is the vertical distance of the heat wave from the top of the bed in the jth section (i.e. the thickness of the burnt material), the time remaining at burnthrough is given by

$$\theta_{b_j} = \frac{h - z_j}{u_j} = \frac{(A - Bd_{p_j})(h - z_j)}{h} \tag{1.52}$$

Also, if x_j is the distance which the jth section is required to travel in time θ_j to ensure that the burnthrough position is correct, the required horizontal velocity for the jth section is given by

$$v_j = \frac{x_j}{\theta_{b_j}} = \frac{x_j h}{(A - Bd_{p_j})(h - z_j)} \tag{1.53}$$

There is, therefore, an ideal velocity for each section of the strand which would be one and the same value if the mean particle diameter did not vary between sections. Since the diameter does vary, the values of the ideal velocities, $v_1, v_2, \ldots, v_j, \ldots, v_N$ are different. At a given instant of time, however, all sections must move at the same velocity. The question that arises, therefore, is how to determine a control law for strand speed which will keep the burnthrough point as close as possible to the ideal position; or better still, minimize the wastage arising from undersintered and oversintered material.

1.5.2 Factors affecting control strategy

An apparently simple strategy that might be adopted would be to calculate $v_1 \cdots v_N$ and drive the strand at the average speed v_a where

$$v_a = \frac{1}{N}(v_1 + v_2 + \cdots + v_N) \tag{1.54}$$

Alternatively, a better result may be achieved by averaging the ideal velocities for those sections nearer to the output end of the strand, say from $v_{N/2}$ to v_N or $v_{3N/4}$ to v_N.

In a section where all the material becomes sintered before the section reaches the correct burnthrough position, the calculated speed for that section would be infinite. Also, for a section in which some material remains unsintered at the correct burnthrough position, the calculated speed for that section is zero. It follows, therefore, that if the calculated 'average' speed is to be meaningful, limits must be placed on the maximum and minimum speed which any one section may contribute to the calculation, i.e. $v_{\min} < v_j < v_{\max}$.

Other factors affecting the control strategy and its effectiveness are:

(a) The frequency with which the disturbances may be measured. Data available for these investigations gave a new value for d_p at 100-second intervals, which means that only ten data points for d_p are on the strand at any given time. However, in practice, the values of d_p vary only gradually and a change of value every 100 seconds is probably sufficient to represent the input disturbance.

(b) The update frequency of strand speed v_a. Each period of control action is taken here as 20 seconds. This should be sufficiently short to have an overall effect approximating to continuous control and yet not so short that the time constant of the speed control system need be considered.

(c) The imposition of limits on the calculated required speed v_j for each bed section has already been mentioned. It is equally important that the averaged demand v_a should be practically realizable, e.g. a system involving step-changes in speed of, say, $\pm 20\%$ every 20 seconds would clearly be unrealistic.

(d) The number of sections assumed in modelling the bed; the larger the number of sections, the more accurate the model will be. 5000 sections are assumed here.

The effectiveness of different strategies is judged on the basis of the marginal strand utilization, i.e. the ratio of increased utilization due to using a particular control strategy to the utilization in the uncontrolled case, expressed as a percentage. Utilization is quantified by computing percentage wastage, i.e. both undersintered and oversintered material. The percentage of undersintered material—more precisely, unsintered material—is obtained by computing the ratio: (height of the front edge of the heat wave at the correct burnthrough position) ÷ (bed height or thickness) × 100. The oversintered material is computed by assuming that the heat wave continues to burn downwards below the bottom of the bed at the same rate as it burned within the bed. This yields, at the correct burnthrough position, a negative 'height' of the heat wave which again can be expressed as a percentage of the bed thickness. The total percentage wastage of strand output over any given period is obtained by accumulating both undersintered and oversintered percentages for that period.

1.5.3 Performance of the controlled strand
The input disturbance in the form of pellet diameter (d_p) values was obtained from a typical record of plant data. The simulation run was first carried out for an 'uncontrolled' strand; that is one which, in the absence of input disturbances, would operate at a constant speed v_c designed to give the correct burnthrough point with a pellet diameter midway between the maximum and minimum values obtained from plant data. The uncontrolled strand is then subjected to typical pellet diameter disturbances over a total period of 9000

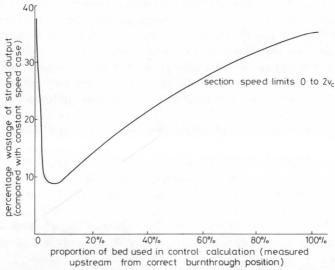

Fig. 1.23 *Wastage of material as a function of the proportion of bed length used in control calculation*

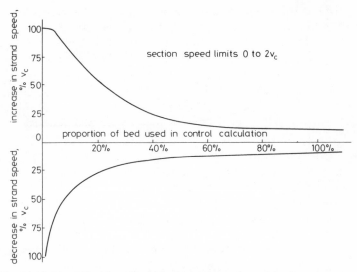

Fig. 1.24 *Maximum and minimum values of strand speed as a function of the proportion of bed length used in control calculation*

seconds (i.e. 9 × the transit time). Information obtained from this case is used as a basis by which the performance of various control systems can be assessed.

The effectiveness of the control strategy outlined in the previous section is shown in Fig. 1.23. For this case $v_{min} = 0$ and $v_{max} = 2v_c$. The number of sections used in the calculation of v_a is varied and is simply represented in the diagram as a distance measured back from the correct burnthrough position. Figure 1.23 shows that by using the prescribed strategy and calculating v_a over the whole length (5000 sections), a reduction in wastage to 35% of the wastage produced in the uncontrolled case is achieved. As the number of sections is reduced, i.e. the calculation of v_a depends more upon the pellet diameters in the sections nearer to the correct burnthrough position, so the reduction in wastage improves to about 10%. The best results are obtained if the averaging procedure is confined to roughly 10% of the bed. However, this does depend, to some extent, on the limits v_{min} and v_{max} imposed in calculating strand speed. The relatively large wastage occurring when less than 5% of the bed length is used in the control calculation is due to the fact that the large speed demands which are required to give accurate burnthrough cannot be met in practice. In these investigations strand speed is limited to a value of $2v_c$. The maximum and minimum values of strand speed v_a calculated for a whole range of lengths are shown in Fig. 1.24. Large changes in strand speed occur if the averaging procedure is carried out using a small proportion of strand length. The rapid switching of speed by large amounts is clearly impractical and it is probably unrealistic to think in terms of using less than

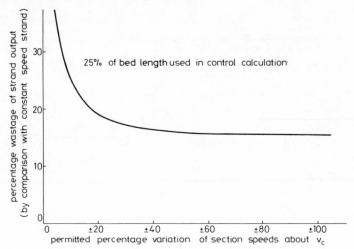

Fig. 1.25   *Effect of limiting the section speed in control calculation*

25% (12·5 m) of the strand for the averaging calculation. There must be a trade-off between process efficiency and controller feasibility.

The limits placed on the maximum and minimum speeds for each section (v_{max} and v_{min}) must also affect control system efficiency. Figure 1.25 shows the percentage wastage (compared with the uncontrolled case) as a function of these limits. The case in which 25% of the strand is used for the averaging process is chosen for this demonstration. Although the wastage decreases as the limits are widened, the curve levels off for limits in excess of about ±30%

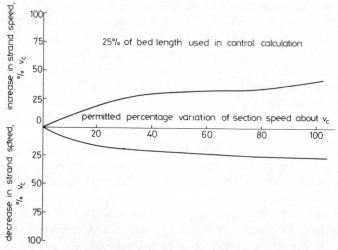

Fig. 1.26   *Maximum and minimum strand speeds demanded when section speeds are limited in control calculation*

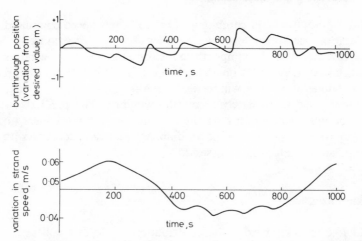

Fig. 1.27 *Process response over a period of 1000 seconds using the 'optimum' control strategy*

of the mean speed. The effect of changing the values of v_{max} and v_{min}, the section-velocity limits, is reflected in the actual strand speed demanded and Fig. 1.26 shows this interrelationship. The $\pm 30\%$ limits on section velocity correspond to a range of $+26\%$ to -18% variation in strand speed. The corresponding reduction in wastage (compared with the uncontrolled case) is a significant 82%. Furthermore, the maximum change in control settings for each controller update period was found to be only 4%. The control strategy therefore appears to be feasible. Figure 1.27 shows the dynamic behaviour of burnthrough position and the corresponding speed adjustment. The burnthrough point is seen to vary by less than 0·7 m from the correct position using the above control strategy. In the uncontrolled case the maximum deviation is 3·6 m.

1.5.4 Comparison with feedback control
The method used in the above control scheme involves the storing of information (effective pellet diameter) obtained from measurements taken at the beginning of the strand and the continual use of a model to predict the strand speed required by each section as it progresses along the strand. In theory, the method is equivalent to a scheme in which the speed required by each section is estimated from measurements actually taken at points along the strand. Thus, the results obtained using the final 5% of the strand for the calculation of v_a are equally representative of the case in which a feedback signal, obtained by sensing the temperature in the final windbox, is used to manipulate strand speed. It may further be argued that the good results obtained by averaging v_j for the 25% of the bed adjacent to the burnthrough point could equally well be produced by measuring windbox temperatures in the final

25% of the windboxes and combining these to form a speed command signal. Indeed, at least one known form of controller uses the temperature in the final windbox together with the temperature measured in an earlier windbox. Such a system, however, depends on there being a good correlation between the measured temperature in a windbox and the height of the heat wave in the section of the bed immediately above the windbox. The predictive scheme, however, is equally dependent on the reliability of the relationship between burnthrough time and effective pellet diameter and the accuracy with which effective pellet diameter can be assessed.

1.6 Acknowledgments

I should like to acknowledge the contributions of former colleagues (particularly C. E. Carter and I. R. Dash) towards the work which has formed the basis of this chapter; also the Science Research Council for financially supporting part of the work and the British Steel Corporation for providing access to data used in the simulation exercises.

1.7 References

1 BALL, D. F., DARTNELL, J., DAVISON, J., GRIEVE, A., and WILD, R.: 'Agglomeration of iron ores' (Heineman Educational Books, 1973)
2 BOUCRAT, M., and ROCHAS, R.: 'Sintering of iron ore on grates', Rev. Met., 1968, 65(12), pp. 835–884 (in French)
3 BRAGARD, A.: 'Control and automation of the sintering process', Ironmaking Tomorrow, ISI publication 102, 1967, pp. 71–83
4 CARMAN, P. C.: 'Fluid flow through packed beds', Trans. AIChE, 1937, 15, pp. 150–163
5 CARTER, C. E., and ROSE, E.: 'A matched model of an iron ore sinter strand process assuming transitional air flow conditions', Proc. Symp. IMC on Measurement and control of quality in the steel industry, Sheffield, 1974, pp. 10/1–10/10
6 CARTER, C. E., and ROSE, E.: 'The simulation of an iron ore sinter strand process', Proc. UKAC Fifth Control Convention, Bath, 1973, pp. 77–85
7 DASH, I., CARTER, C. E., and ROSE, E.: 'Heat wave propagation through a sinter bed; a critical appraisal of mathematical representations', Proc. Symp. IMC on Measurement and control of quality in the steel industry, Sheffield, 1974, pp. 8/1–8/7
8 DASH, I., and ROSE, E.: 'An analytical study of processes occurring in an iron ore sinter bed', Proc. Second IFAC Symp. on Automation in mining, mineral and metal processing, Johannesburgh, 1976, T9B2, pp. 1–11
9 DASH, I. R., and ROSE, E.: 'Sinter strand control; a simulation investigation', IFAC Seventh World Congress, Helsinki, 1978, pp. 151–158
10 EFIMENKO, G. G., KOVALEV, D. A., SULIMENKO, E. I., KNYAZHANSKII, M. M., and ISPOLATOV, V. B.: 'Production of higher-strength fluxed pellets', Steel in the USSR, 1978, 8, pp. 123–125
11 ELLIOTT, J. F.: 'Some problems in macroscopic transport', Trans. Metall. Soc. AIME, 1963, 227, pp. 802–820
12 ERGUN, S.: 'Fluid flow through packed columns', Chem. Eng. Prog., 1952, 48, pp. 89–94

13 GRIEVE, A.: 'Pilot plant work in the field of iron ore agglomeration', ISI special report 96, Pilot plants in the iron and steel industry, 1966, pp. 175–183

14 KINSEY, F. W.: 'Design features of a modern sinter plant', *Proc. AIME Ironmaking Conf.*, Detroit, 1970, pp. 97–106

15 KOH, J. C., DUTTON, J. L., BENSON, B. A., and FORTINI, A.: 'Friction factor for isothermal and non-isothermal flow through porous media', *Trans. Am. Soc. Mech. Eng., J. Heat Transfer*, 1977, **99**(3), pp. 367–373

16 MUCHI, I., and HIGUCHI, J.: 'Theoretical analysis of sintering operation', *Trans. ISI Jpn*, 1972, **12**, pp. 54–63

17 NYQUIST, O.: 'Effects of lime on the sintering of pure hematite and magnetite concentrates', *International Symposium on Agglomeration, Philadelphia*, 1961, pp. 809–858

18 OGG, A. F., and JENNINGS, R. F.: 'The economics of sintering', *Ironmaking and Steelmaking*, 1977, **3**, pp. 153–158

19 PARKER, A. S., and HOTTEL, H. C.: 'Combustion rate of carbon', *Ind. & Eng. Chem.*, 1936, **28**, p. 1334

20 PENGELLY, A. E., and CARTER, G. A.: 'Automatic control systems in steelplants: a user's view', *Ironmaking and Steelmaking*, 1978, **6**, pp. 273–281

21 PERRY, R. H., and CHILTON, C. H.: 'Chemical engineers handbook', 4th ed. (McGraw Hill, 1968)

22 POOS, A., MEUNIER, G., and LUCKERS, J.: 'Automation of sinter plant', ISI special report 152, 1973, pp. 139–146

23 PRICE, C., and WASSE, D.: 'Relationship between sinter chemistry, mineralogy and quality and its importance in burden optimisation', ISI publication 152, 1973, pp. 32–52

24 PRIGORINE, I., and DEFAY, R.: 'Chemical thermodynamics' (Longmans, 1962)

25 ROSE, E., BROWN, J. C., and CARTER, C. E.: 'Lumped-parameter models for control of an iron-ore sinter strand process', *Proc. IFAC Sixth World Congress, Boston*, 1975, **39**(5), pp. 1–10

26 ROSENBROCK, H. H., and STOREY, C.: 'Computational techniques for chemical engineers' (Pergamon Press, 1966)

27 SAUNDERS, O. A., and FORD, H.: 'Heat transfer in the flow of gas through a bed of solid particles', *J. Iron Steel Inst.*, **141**, 1940, pp. 291–328

28 SCHLUTER, R., and BISTRANES, G.: 'The combustion zone in the iron ore sintering process', *International Symposium on Agglomeration, Philadelphia*, 1961, pp. 585–637

29 SZEKELY, J., and CARR, G.: 'Non-isothermal flow of gases through packed beds', *Trans. Metall. Soc. AIME*, 1968, **242**, pp. 918–921

30 VISVANATHAN, S.: 'Sinter versus pellets', *TISCO*, 1970, **17**(4), pp. 111–114

31 VOICE, E. W., BROOKS, S. H., DAVIES, W., and ROBERTSON, B. L.: 'Factors controlling the rate of sinter production', *ISI special report 53*, 1955, pp. 43–98

32 VOSKAMP, J. H., and BRASZ, J.: 'Digital simulation of the steady behaviour of moving bed process', *Meas. and Control*, 1975, **8**, pp. 23–32

33 YOUNG, R. W.: 'A dynamic simulation of the induration process for pelletising ores', *Proc. UKSC Conf. Comput. Simulation, Chester*, 1978, pp. 119–128

34 YOUNG, R. W.: 'Dynamic mathematical model of sintering process', *Ironmaking and Steelmaking*, 1977, **6**, pp. 321–328

Ironmaking and steelmaking—II: Modelling and control of steelmaking in the A.O.D. process

K. V. Fernando and H. Nicholson

List of principal symbols

a, b parameters associated with the equilibrium constant K

d, e, f parameters associated with the function γ_C

A activity of the chemical species indicated by subscript

C_p specific heat of the molten metal

h_a, h_c, h_r parameters associated with heat losses

H enthalpy change due to oxidation of the species indicated by subscript

J_p, J_l the rates of production and loss of heat

K equilibrium constant

M molecular weight indicated by subscript

P the total pressure in the melt

P_{CO} the partial pressure of CO

p mass of carbon oxidized by one unit volume of oxygen

q mass of chromium oxidized by one unit volume of oxygen

r total volume rate of injection of gases

r_{CO} rate of production of CO

t time

T temperature of the bath

T_e ambient temperature

u volume rate of injection of oxygen

v volume rate of injection of argon

w mass of molten metal

x mass of carbon in the melt

y mass of chromium in the melt

α empirical attenuation parameter

γ_C activity coefficient of carbon

List of principal chemical symbols

Ar   argon
C   carbon
CO   carbon monoxide
Cr   chromium
Cr_mO_n chromium oxide

2.1 Introduction

Mathematical modelling of a system is a science as well as an art. Much skill, experience and care are required to obtain a fairly representative set of mathematical relations to describe the actual process, especially if it does not conform to simple laws of nature.

In metallurgical processes such as argon oxygen decarburization (AOD) of steel,[4-13] complex reactions and violent physical changes occur which themselves are governed by subprocesses. Analytical modelling of such systems are difficult if not impossible. Even if a formal set of mathematical relations is derived using laws of chemistry, problems arise if a numerical solution or a simulation of the model is required. It is not easy to gauge or to decide how much emphasis should be given to the different subprocesses and disagreements exist between experts. Further, parameters of the system should be known correctly which otherwise degrade the analytical accuracy and these are not always easy to compute or measure. Hence, quasi-analytical modelling is the standard practice for such involved processes.

Complex models do not always give better results and often simple models offer more reliable information. It is easy to analyse qualitative aspects of a simple model which otherwise would be hidden in complex equations.

Most metallurgical models are based on a 'stirred tank' approach where average conditions are assumed through the reactor. Though such assumptions are not exact, lumped parameter models normally give acceptable results.

In this study a new simple model of the AOD process is developed from basic chemical principles. It is shown that this model can be employed to design a controller whose structural simplicity allows the possibility of feedback control.

2.2 The process and desired model structure[4-13]

Due to the fierceness of the furnace flame and the nature of the carbon-chromium-iron equilibria it is not possible to produce stainless steels of low carbon levels in an electric arc furnace. This led to the invention of the AOD process by the Union Carbide Corporation in the 1950s and subsequent commercial production of stainless steels began in the late 1960s. This process has revolutionized the manufacture of stainless steels and high quality steels can now be produced using this method.

The molten metal is first prepared in a conventional arc furnace and is then transferred to the AOD vessel for decarburization. Oxidation of carbon, silicon and other undesirable elements are achieved by blowing a mixture of oxygen and argon from one or more openings or tuyeres located at the base of the vessel which are immersed in the molten metal when the vessel is in the upright position. A schematic view of an AOD vessel is shown in Fig. 2.1.

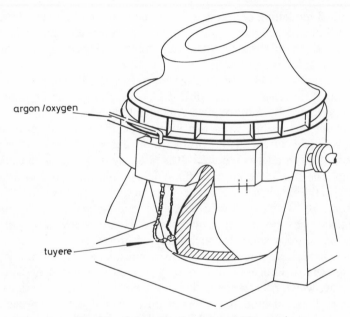

Fig. 2.1 *Schematic view of an AOD vessel*

The prime reason for injecting argon along with oxygen into the AOD process is that this inert gas dilutes the evolved carbon monoxide and thereby decreases the rate of oxidation of chromium.[4] Oxidation of this expensive metal is the main undesired side-effect and argon has the function of discouraging oxygen which otherwise would combine with chromium. The actual chemical kinetics are complex but if a 'black-box' approach is used, it could be said that argon shields chromium from oxidation.

Both carbon and chromium compete for oxygen to be oxidized, and in that respect this chemical system is similar to systems of biological species competing for common sources of food. It has been found in population biology[15, 22] that linear models are inadequate to describe the dynamics and Volterra-type systems and variable structure systems are required. The most simple model of this kind is bilinear with respect to states and controls. The controllers in this process are the rates of injection of oxygen and argon.

One of the basic disadvantages of Fruehan's[7] model of the process from the point of view of control theory, is that the injection rates of gases do not appear linearly. These nonlinearities do not complicate simulation studies but standard methods for designing controllers are not known at present. Therefore, if possible, the controller should occur linearly. Such systems, which are otherwise nonlinear are being studied increasingly.[14, 21]

2.2.1 The model
2.2.1.1 Oxidation process. The model is based on the assumption that the injected oxygen primarily oxidizes the chromium in the neighbourhood of the tuyere zone, and the chromium oxide formed oxidizes the carbon as it rises in the bath with the argon bubbles. It is further assumed that the oxidation of carbon by chromium oxide is the rate-controlling reaction.

The above postulates are similar to those of Fruehan,[7] and the main difference is that it is initially assumed that this rate-controlling reaction achieves its equilibrium, and hence the rate of oxidation of carbon can be computed using the equilibrium constant of the reaction. Fruehan[7] assumed that this reaction does not attain equilibrium and that the rate of reaction can be described by an empirical first-order diffusion-type differential equation. In the present study, any departure from equilibrium will be assumed to be accountable for by the introduction of an empirical (attenuation) constant.

The injected oxygen oxidizes the chromium as given by the following reaction

$$2m\mathrm{Cr} + n\mathrm{O}_2 = 2\mathrm{Cr}_m\mathrm{O}_n$$

There is some controversy[1] over the form of the oxide and it will be assumed that this is $\mathrm{Cr}_2\mathrm{O}_3$. As the chromium oxide particles move away from the tuyere zone with argon bubbles, the following reaction which governs the process, occurs

$$1/n\mathrm{Cr}_m\mathrm{O}_n + \mathrm{C} = m/n\mathrm{Cr} + \mathrm{CO}$$

The equilibrium of the reaction is determined by the activities of the reactants related by the equilibrium constant given by

$$K = \frac{(A_{\mathrm{Cr}})^{m/n} P_{\mathrm{CO}}}{\gamma_{\mathrm{C}} A_{\mathrm{C}} (A_{\mathrm{Cr}_m\mathrm{O}_n})^{1/n}} \tag{2.1}$$

The equilibrium constant K is temperature-dependent and can be written in the form given by

$$K = \exp(a - b/T) \qquad a > 0, b > 0$$

The activities of carbon and chromium are approximately equal to their

MDS - E

molar fractions in respect to the total mass of liquid metal, that is,

$$A_C \simeq (x/M_C)/(w/M_{Fe})$$
$$A_{Cr} \simeq (y/M_{Cr})/(w/M_{Fe})$$

(2.2)

where x and y are respectively the masses of carbon and chrome dissolved in the bath. The activity of Cr_mO_n can be taken as unity. The function γ_C which relates activity of carbon in chromium to that of graphite can be written in the form[2]

$$\exp \gamma_C = -dA_{Cr} + eA_C - f \qquad d, e, f > 0$$

The partial pressure of carbon monoxide in the molten metal is given by

$$P_{CO} = r_{CO}P/(r_{CO} + v)$$

(2.3)

The rate of production of carbon monoxide r_{CO} can be calculated from the following reaction

$$2C + O_2 = 2CO$$

then

$$r_{CO} = -\frac{dx}{dt}\bigg/p = -\dot{x}/p$$

(2.4)

Combining Eqns. (2.1)–(2.4) and rearranging terms gives the differential equation which governs the oxidation of carbon as

$$\dot{x} = -p\bar{F}v$$

where

$$\bar{F} = 0{\cdot}5\gamma_C Kx/(By^{m/n} - \gamma_C Kx)$$
$$B = P(M_{Fe}/wM_{Cr})^{m/n}/(M_{Fe}/wM_C)$$

The above equation which gives the rate of decarburization is derived assuming that the reaction is in equilibrium. However, the actual rate of oxidation may be less than the rate predicted by the equilibrium constant, and hence an empirical attenuation constant has to be included.

Most of the secondary effects which were not taken into account in the model can be lumped into this empirical parameter α, and in that respect, it is similar to the empirical parameter α used by Fruehan[7] and also by Roy et al.[12, 13] This parameter is combined with the function $\bar{F}$ to give $F = \alpha\bar{F}$.

At high carbon levels the value of F can be very high and the reaction is not at equilibrium. There is insufficient oxygen to oxidize the excess carbon and the rate of decarburization is constrained by the rate of injection of oxygen. Assuming that the bulk of oxygen reacts only with carbon at high levels, the maximum rate of decarburization is given by

$$\max(-\dot{x}) = pu$$

The following differential equations can be obtained once the constraint is incorporated

$$\dot{x} = -pFv \qquad 0 \le F \le u/v$$
$$\dot{x} = -pu \qquad F < 0 \text{ or } F > u/v \tag{2.5}$$

As silicon and manganese burn off at the beginning of the process, the oxygen demand by these elements at later stages of the process is small but not entirely negligible. Some oxygen dissolves in the molten metal and some oxidizes the iron. However, the bulk of it oxidizes the prime active constituents carbon and chromium, and the oxygen balance for the process yields

$$-\dot{x}/p - \dot{y}/q = u$$

The governing equations for oxidation of chromium are given by

$$\dot{y} = -qu + qFv \qquad 0 \le F \le u/v$$
$$\dot{y} = 0 \qquad 0 > F \text{ or } F > u/v$$

The oxidized carbon leaves the bath as carbon monoxide and chromium enters the slag as oxide. The mass of the molten metal which can be calculated from the following differential equation

$$\dot{w} = \dot{x} + \dot{y}$$

is required to calculate the percentages and activities of the elements.

2.2.1.2 The heat balance. If the heat generated by oxidation of elements such as silicon and manganese is neglected, since it is assumed that this will be generated very early in the process, the rate of production of heat is given by

$$J_p = (-\dot{x})(-H_C) + (-\dot{y})(-H_{Cr})$$

At high temperatures the heat loss is dominated by radiation but there is some conduction loss and also some of it is carried away by argon and carbon monoxide. If the heat loss to carbon monoxide is neglected, the approximate total loss is given by

$$J_l = h_c(T - T_e) + h_r(T^4 - T_e^4) + h_a v(T - T_e)$$

The first term represents the heat loss due to conduction and the second that due to radiation. The last term is due to the sensible heat increase of argon. The coefficients h_c and h_r depend upon the geometry of the vessel and h_a is the specific heat of argon per unit volume.

The total heat capacity of the vessel is equal to the sum of heat capacities of the molten metal, the slag and the refractories. As an approximation, it can be taken as the product of the mass of the molten metal and its specific heat. The net rate of heat production is equal to the rate at which heat is accumulated.

The governing equation for the temperature is thus given by

$$\dot{T} = (J_p - J_l)/(WC_p)$$

2.2.1.3 The structure of the derived model. Except at the first phase of the process, when only the excess carbon burns, the oxidation of chromium and carbon is described by the following differential equations, as derived in the last section

$$\begin{pmatrix} \dot{x} \\ \dot{y} \end{pmatrix} = \begin{pmatrix} -pF \\ qF \end{pmatrix} v + \begin{pmatrix} 0 \\ -q \end{pmatrix} u \qquad 0 \le F \le u/v$$

This system is linear in control but nonlinear in the state, and this linearity will be exploited in designing a suitable controller. The function of argon in this process is to protect the chromium from oxidation which is explicitly evident from the structure of the system equations, and it is present as a first-order effect. This is not so in other published models.[7, 12, 13]

The structure of the model has features common with models of biological competing species. If compared with such ecological systems,[17–19] the controls could be considered as rates of harvesting (fishing, culling) or enrichment of species.

When the carbon level is low (at the final stages of the process), the model of Fruehan breaks down due to absence of the controls. Such total invalidation does not occur in the present model. Still, the model could not be expected to give very reliable results at very low concentrations of carbon. The equilibrium constant for the process can be written in the form

$$K = (\bar{A}_{Cr} P_{CO})/(\gamma_C A_C)$$

where $\bar{A}_{Cr} = A_{Cr}^{m/n}$. If the nonlinearity due to the effect of γ_C is neglected, the equilibrium constant is based on proportionalities of the active reactants. In population biology[22] such models are sometimes known as 'primitive statistical models' and they are not considered as valid near the extinction of the species.[23]

2.3 Phase-plane analysis

If the mass of the molten metal W and the nonlinear function γ_C are taken as constant (in a local region), phase-plane analysis can be carried out to investigate the behaviour of the model. The nonlinearity of the model is solely due to the function F, given by

$$F = \alpha \, 0{\cdot}5\gamma_C Kx/(By^{m/n} - \gamma_C Kx)$$

$$= \alpha \, 0{\cdot}5z/(1 - z)$$

where $z = \gamma_C Kx/(By^{m/n})$

A line singularity exists when $z = 1$, but it occurs at very high levels of carbon when only excess carbon burns. Hence, in the region of interest $(0 \le F \le u/v)$, the function is well behaved.

If the rate of injection of argon v is constant, then loci of constant F define constant decarburization rates. These can be found using the relation

$$y^{m/n} = DKx\gamma_C$$

where $D = (\alpha + 2F)/(2FB)$

2.4 Design of an optimal controller

The AOD process was invented to cut down or minimize the oxidation of chromium during decarburization and this should be the primary goal in designing a controller. Secondary costs could incur due to any excess argon which has to be used because of the action of the controller. However, because of the uncertainties and limitations of the model, incorporation of these additional penalties would not be very meaningful. Furthermore, previous studies[16] have demonstrated that the chromium costs dominate. The cost function for minimization is taken as

$$I = \min (y_0 - y_r)$$

where y_0 and y_r are the initial and final values of chromium.

Since the process is not oscillatory and the actual amounts of carbon and chromium monotonically decrease with time, it is possible to design a simple controller. An optimal-aim controller[20] could be obtained by static optimization at each instant of time.

At the beginning of blow, no oxidation of chromium occurs and the maximum practicable oxygen/argon ratio should be used. But when the value F is less than the ratio u/v, the rate of oxidation of chromium is given by

$$\dot{y} = -qu + qFv$$

According to the oxygen balance, the above oxidation is due to the excess of oxygen which failed to decarburize. Oxidation of chromium could be completely avoided if the exact amount of oxygen required is injected. The necessary condition for such zero oxidation is given by

$$\dot{y} = 0$$

and hence the optimal ratio of oxygen to argon at each instant of time is given by

$$u/v = F \tag{2.6}$$

The actual injection rates of gases can be calculated by knowing the total rate of gas injection possible.

If this controller is implemented, the minimized cost function will be zero, resulting in a dynamically optimal solution. In general, optimal-aim controllers are not dynamically optimal and hence this is a special case. If the effects due to temperature variations are neglected, this controller could be considered as time-optimal as well, since it utilizes all available oxygen for decarburization. The optimal controller attempts to cut off completely the oxidation of chromium. This is not possible in practice and this optimal ratio would give a very conservative rate of injection of oxygen. Therefore, this should be considered as the minimum limit of oxygen required at any given instant. Further, this optimal controller could be used as a norm to assess the performance of other controllers.

2.5 Numerical results

Simulation studies of a process have been made employing the developed model and controller. These studies concentrate on that part of the process cycle when $0 \leq F \leq u/v$, (see Section 2.2.1.1). That is, the initial period during which the rate of decarburization and temperature rise is determined by the oxygen injection rate and the concentrations of secondary elements such as silicon have been ignored. This is reasonable since it is common practice to estimate, from experience, the end of this period and to stop the refining process to take a sample for chemical analysis, hence initial conditions for a simulation of the later stages are available.

The object of these studies is to illustrate the sensitivity of the model to the empirical and physical parameters, and the temperature and weight variations have been neglected unless specified. The data employed in these studies are presented in Table 2.1.

2.5.1 The effect of the parameter α (see Fig. 2.2)
The state equation for carbon can be written in the form

$$\dot{x} = -\alpha G x v$$

where $G = F/(x\alpha)$ is approximately constant locally. The value $(-\alpha G v)$ can, therefore, be considered as the eigenvalue of the system and different α's give different time constants for decaying carbon, which can be seen from the trajectories.

2.5.2 The influence of the interaction coefficient γ_C (see Fig. 2.3)
The results presented in Fig. 2.3 demonstrate that the model is not significantly influenced by allowing γ_C to vary and that any variation in response due to γ_C can be absorbed in the parameter α.

Table 2.1

Simulation data		

Initial conditions		

Carbon	x	0·41%
Chromium	y	16·90%
Mass of metal	w	42000 kg
Temperature	T	1650°C

Gas injection rates (controls)		

		Stage 2	Stage 3
Duration	t	20 min	13 min
Argon rate	v	10·4 m^3 min^{-1}	21·5 m^3 min^{-1}
Oxygen rate	u	20·8 m^3 min^{-1}	11·1 m^3 min^{-1}
Maximum total	r	33·0 m^3 min^{-1}	33·0 m^3 min^{-1}

Data for the temperature model		

Ambient temperature	T_e	100°C
Coefficient (conduction loss)	h_c	0
Coefficient (radiation loss)	h_r	0·5 10^{-14} kcal (K)$^{-4}$ min^{-1}
Coefficient (argon loss)	h_a	0
Specific heat	C_p	0·16 kcal kg^{-1} (K)$^{-1}$
Enthalpy	H_C	$-33,350$ kcal (kg mol C)$^{-1}$
Enthalpy	H_{Cr}	$-214,600$ kcal (kg mol 2 Cr)$^{-1}$

Other data	

Equilibrium constant	$K(T) = \exp(62\cdot884 - 94337/T)$
Activity coefficient	$\gamma_C = \exp(2\cdot302585(-0\cdot25 - 2\cdot4A_{Cr} + 22\cdot0A_C))$
Total pressure	$P = 1000$ mbar

2.5.3 The effect of weight variations (see Fig. 2.4)
It can be seen from the responses presented in Fig. 2.4 that the dynamics of the process as modelled are not sensitive to the effects of weight variation. The only influence to be seen is the anticipated change in the weight percentage concentration of chromium.

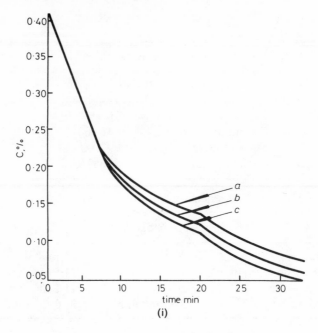

(i)

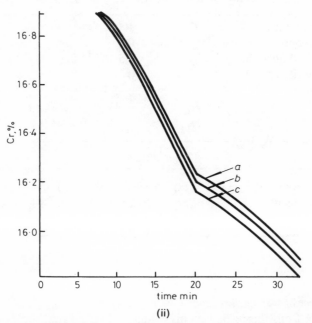

(ii)

Fig. 2.2   *The effect of* α (i) *Carbon response* (ii) *Chrome response*
   a α = 0·6
   b α = 0·8
   c α = 1·0

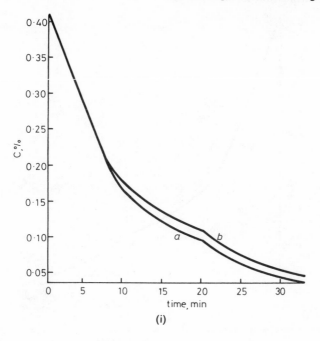

(i)

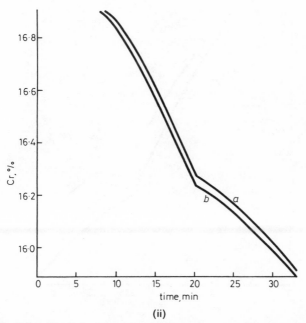

(ii)

Fig. 2.3 *The influence of the interaction coefficient* γ_C (i) *Carbon response* (ii) *Chrome response*
a γ_C varying
b γ_C constant

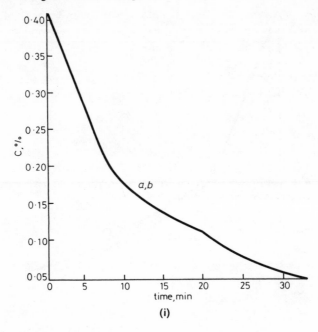

(i)

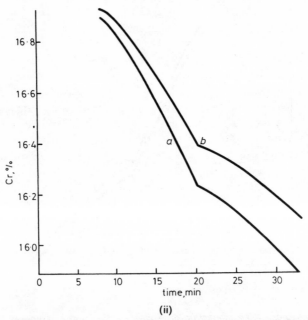

(ii)

Fig. 2.4 *The effect of weight variations* (i) *Carbon response* (ii) *Chrome response*
a w varying
b w constant

2.5.4 The effect of temperature on process response (see Figs. 2.5, 2.6)
The temperature response presented in Fig. 2.5(i) is similar in terms of the net change in temperature during the second and third intervals of the refining cycle, to the limited thermal data available for the process. The effect of the increasing process temperature is seen to be a retardation of the oxidation of chromium which is to be expected from the chemistry of the process.

2.5.5 Control strategies (see Figs. 2.7, 2.8)
The oxygen and argon injection strategies presented in Fig. 2.7 demonstrate the extent to which the optimal injection strategy required to minimize chromium losses (see Fig. 2.8(ii)) differs from the conventional switched multilevel control.[7, 10] It is also apparent from the carbon trajectory (see Fig. 2.8) that the optimal control also gives a reduction in operating time.

2.6 Conclusions

A new nonlinear dynamic model of the AOD process has been presented. The structure of the model has been analysed to show that the dynamic performance of the process has similarities to the dynamics of processes currently the subject of much research interest in the general field of systems theory. Furthermore, it has been shown that since the process model is linear in control, even though it is nonlinear in state, a simple control law to minimize chrome losses can be obtained. Simulation studies using this model have revealed that the process dynamics are not significantly sensitive to the effects of interaction in chemical activities and weight variations. Indeed the results presented in Section 2.5 demonstrate that any uncertainty in dynamic behaviour due to mild inaccuracy in the model parameters can be absorbed into the empirical parameter α.

The simple structure of the control law proposed, see Eqn. (2.6), is of particular practical interest in view of the current industrial efforts to implement waste-gas analysis equipment to measure rates of decarburization. Such equipment would provide a measurement of the volume flow rate of carbon monoxide, hence from Eqn. (2.5) it is seen that the quantity Fv would be available. The practical implementation of a feedback control system based on such a measurement would require the use of a well-tuned filter to obtain estimates of the states, and work on the design of optimal filters to be used in such conditions has been conducted.[3] The remainder of the control system would consist of standard ratio controllers used with re-heat furnaces.

2.7 Acknowledgment

The authors wish to acknowledge the work of Dr. F. M. Boland of Trinity College, Dublin in initiating this project while he was at the Department of Control Engineering, University of Sheffield.

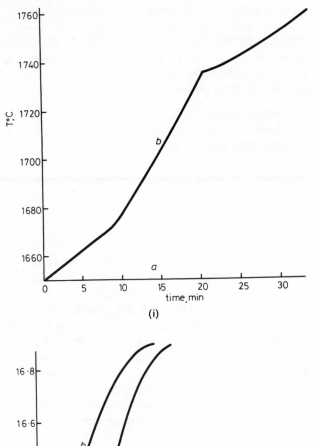

(i)

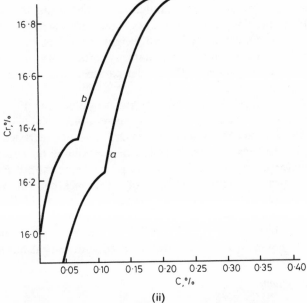

(ii)

Fig. 2.5 *Temperature effects* (i) *Temperature profile* (ii) *Chrome-carbon phase-plane response*
  a Without temperature model
  b With temperature model

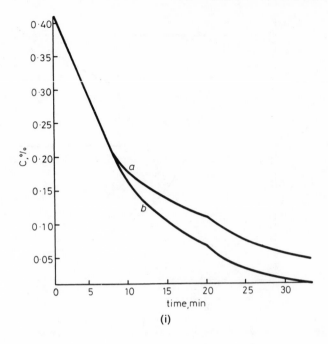

(i)

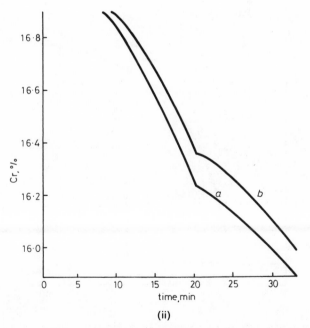

(ii)

Fig. 2.6   *Temperature effects* (i) *Carbon response* (ii) *Chrome response*
   a Without temperature model
   b With temperature model

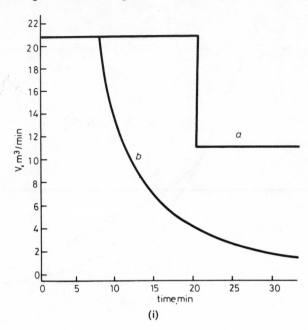

(i)

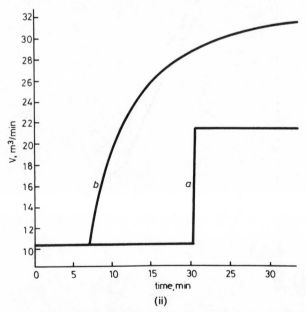

(ii)

Fig. 2.7   *Control strategies* (i) *Oxygen flow rate* (ii) *Argon flow rate*
a Two-level control
b Optimal control

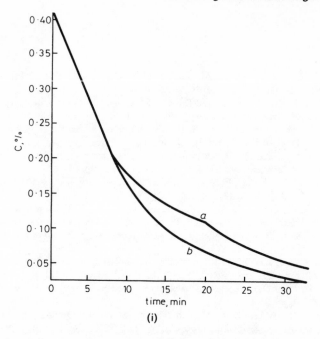

(i)

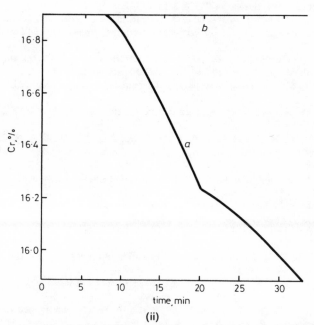

(ii)

Fig. 2.8 *Influence of control strategies* (i) *Carbon response* (ii) *Chrome response*
a Two-level control
b Optimal control

2.8 References

1 BILLINGS, S. A., BOLAND, F. M., and NICHOLSON, H.: 'Electric arc furnace modelling and control', SRC Grant Report B/RG/29402, 1976
2 ELLIOTT, J. F., GLEISER, M., and RAMAKRISHNA, V.: 'Thermochemistry for steelmaking', Vol. II (Addison-Wesley, Reading, Mass., 1963)
3 BOLAND, F. M., and NICHOLSON, H.: 'Estimation of the states during refining in electric arc furnaces', Proc. IEE, 1977, 124(2), pp. 161–166
4 ASAI, S., and SZEKELY, J.: 'Decarburization of stainless steel: Part 1. A mathematical model for laboratory scale results', Metall. Trans., 1974, 5, pp. 651–657
5 CHOULET, R. J., DEATH, F. S., and DOKKEN, R. N.: 'Argon-oxygen refining of stainless steel', Can. Metall. Q., 1971, 10(2), pp. 129–136
6 AUCOTT, R. B., GRAY, D. W., and HOLLAND, G. G.: 'The theory and practice of the argon/oxygen decarburizing process', J. W. Scotland Iron Steel Inst., 1971, 79, pp. 97–127
7 FRUEHAN, R. J.: 'Reaction model for the AOD process', Ironmaking and Steelmaking, 1976, 3, pp. 153–158
8 HILTY, D. C., and KAVENY, T. F.: 'Melting of stainless steels', from 'Handbook of stainless steels', PECKNER, D., and BERNSTEIN, I. M. (Eds.) (McGraw-Hill, N.Y., 1977)
9 HODGE, H. L.: 'AOD process and its eighty-six months of growth', Ironmaking and Steelmaking, 1977, 2, pp. 81–87
10 IRVING, M. R., BOLAND, F. M., and NICHOLSON, H.: 'Optimal control of the argon oxygen decarburising steelmaking process', Proc. IEE, 1979, 126(2), pp. 198–202
11 LEACH, J. C. C., RODGERS, A., and SHEEHAN, G.: 'Operation of the AOD process in BSC', Ironmaking and Steelmaking, 1978, 3, pp. 107–120
12 ROY, T. D., and ROBERTSON, D. G. C.: 'Mathematical model for stainless steelmaking, Part 1: Argon-oxygen and argon-oxygen-steam mixtures', Ironmaking and Steelmaking, 1978, 5, pp. 198–206
13 ROY, T. D., ROBERTSON, D. G. C., and LEACH, J. C. C.: 'Mathematical model for stainless steelmaking, Part 2: Application to AOD heats', ibid., pp. 207–210
14 BROCKETT, R.: 'Nonlinear systems and differential geometry', Proc. IEEE, 1976, 64(1), pp. 61–72
15 PORTER, W. A.: 'An overview of polynomic systems theory', Proc. IEEE, 1976, 64(1), pp. 18–23
16 WOODSIDE, C. M., PAGUREK, B., PAUKSENS, J., and OGALE, A. N.: 'Singular arcs occurring in optimal electric steel refining', IEEE Trans. Autom. Control, 1970, AC-15(5), pp. 549–556
17 BRAUER, F.: 'De-stabilization of predator-prey systems under enrichment', Int. J. Control, 1976, 23(4), pp. 541–552
18 BRAUER, F., and SANCHEZ, D. A.: 'Constant rate population harvesting: Equilibrium and stability', Theor. Population Biol., 1975, 8, pp. 12–30
19 YODZIS, P.: 'The effects of harvesting on competitive systems', Bull. Math. Biol., 1976, 38, pp. 97–109
20 BARNARD, R. D.: 'Continuous-time implementation of optimal-aim controls', IEEE Trans. Autom. Control, 1976, AC-21(3), pp. 432–434
21 HERMES, H., and HAYNES, G.: 'On the nonlinear control problem with control appearing linearly', J. SIAM Control, 1963, 1(2), pp. 85–108
22 GOEL, N. S., MAITRA, S. C., and MONTROLL, E. W.: 'Nonlinear models of interacting populations' (Academic Press, New York, 1971)
23 HIRSCH, M. W., and SMALE, S.: 'Differential equations, dynamical systems, and linear algebra' (Academic Press, New York, 1974)

Chapter 3

Modelling and identification
of a
three-phase electric-arc furnace

S. A. Billings

3.1 Introduction

This chapter is an attempt to illustrate the development of a mathematical model of an industrial process by combining analytical modelling techniques with a system identification study. Although both these approaches can often be applied in isolation to develop a mathematical description of a process, the lack 'of detailed knowledge of the high-power arc discharge suggests a combination of these techniques in the case of the electric-arc furnace.

The objective of the present study is to develop three-phase models of both arc-impedance- and arc-current-controlled electric-arc furnaces. Initially a single-phase model of the electrode position controller, arc discharge and furnace transmission system are developed for an arc-impedance-controlled furnace using analytical modelling techniques and assuming zero interaction between the three regulators. The assumptions and approximations associated with this model are investigated by designing experiments and conducting tests on a production arc furnace. Properties of the arc discharge, interaction between the regulators and a pulse transfer function representation of the electrode position controller are identified. These results are extended and three-phase models of an arc-impedance- and arc-current-controlled electric-arc furnace are derived.

3.2 The electric-arc furnace

Electric-arc furnaces are widely used throughout the world to melt and refine steel. The present study relates to a 135-tonne 35-MVA production arc furnace. The furnace consists of a refractory lined shell with three electrodes which pass through holes in the roof. The roof and electrode structure can be swung aside in a horizontal plane to permit scrap charging from an overhead basket. Electrical power is supplied to the furnace electrodes through busbars and water-cooled flexible cables from the furnace transformer. Heat is transferred to the scrap steel from electric arcs (typically 36 kA at 560 V) drawn between the tips of the electrodes and the metallic charge. A schematic diagram of the arc-impedance-controlled furnace is illustrated in Fig. 3.1.

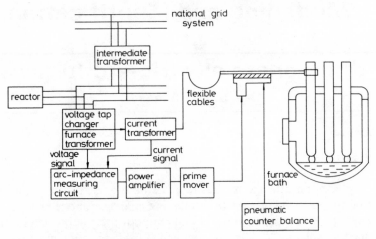

Fig. 3.1 *A schematic diagram of an arc-impedance-controlled furnace*

Throughout the period of a melt the arc length varies erratically due to scrap movement within the furnace and some form of control is required to maintain the desired power input level. The existing control philosophy is based upon both short- .and long-term control policies. The long-term policy consists of manually adjusting the power input to the furnace at various stages of the melt by selection of a suitable transformer voltage tap. Short-term dynamic control is based upon maintaining a preselected constant arc impedance by adjusting the electrode position in response to disturbances. Each electrode is individually positioned by an electrode position controller which attempts to maintain a reference arc impedance compatible with the long-term power input schedule.

3.3 Single-phase model

Although a considerable amount of research has been directed towards improving the melting efficiency of the electric-arc furnace there are many aspects of its operation and control which require further study. Various authors[1, 2, 3] have developed single-phase models of electric-arc furnace control systems but very little work has been directed towards studying the properties of the high-power arc discharge[4, 5] and its effects upon short-term dynamic control.

The servomechanism for positioning the electrodes in the furnace under investigation is typical of many installations and consists of an amplidyne Ward-Leonard regulator operating on an arc-impedance error signal. A circuit diagram of one of the three electrode position controllers is illustrated in Fig. 3.2.

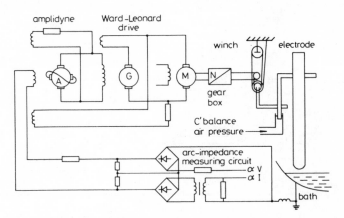

Fig. 3.2   *Ward-Leonard drive electrode-position controller*

The arc-impedance measuring circuit compares currents proportional to transformer secondary voltage and arc current and produces an error signal when they are unequal. Under steady-state conditions the error signal ϵ can be represented by

$$\epsilon = G_5 I_1 - G_4 V_{m1} \tag{3.1}$$

hence the term arc-impedance control. The error signal is applied to the control winding and excites the amplidyne Ward-Leonard drive which positions the electrode through suitable gearing and a winch system.

Interaction between the three electrode regulators is minimized when operating under arc-impedance control. The interaction which arises from properties of the arc discharge may not, however, be negligible although all previous work has assumed this to be the case. Properties of the arc discharge will be investigated in later sections but a single-phase model will be derived initially to provide insight and aid the design of experiments on the furnace.

The Ward-Leonard drive can be modelled using transfer function relationships, derived by analysing equivalent circuits of the electrical machines and can be represented as shown in Fig. 3.3. The defining state-space equations can then be determined as follows,[6] using the symbols defined in Fig. 3.3.

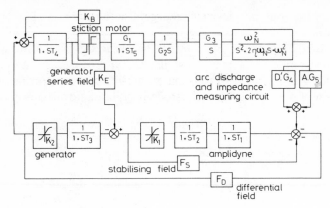

Fig. 3.3   *Single-phase transfer function model*

Amplidyne:

$$\dot{x}_1 = \frac{(-DG_4 - AG_5)x_9 - F_D x_3 - F_S x_2 - x_1}{T_1} \tag{3.2}$$

$$\dot{x}_2 = \frac{k_1 x_1 - x_2}{T_2} \tag{3.3}$$

Generator:

$$\dot{x}_3 = \frac{K_2 x_2 - K_2 K_E x_4 - x_3}{T_3} \tag{3.4}$$

Motor:

$$\dot{x}_4 = \frac{x_3 - F_B x_6 - x_4}{T_4} \tag{3.5}$$

$$\dot{x}_5 = \frac{G_1 x_4 - x_5}{T_5} \tag{3.6}$$

$$\dot{x}_6 = \frac{x_5}{G_2} \tag{3.7}$$

$$\dot{x}_7 = G_3 x_6 \tag{3.8}$$

Mast dynamics:

$$\dot{x}_8 = \omega_N^2(x_7 - x_9) - 2\zeta\omega_N x_3 \tag{3.9}$$

$$\dot{x}_9 = x_8 \tag{3.10}$$

The weight of the electrodes and supporting mast structure is pneumatically counterbalanced and the dynamics of the electrode are represented in the model by a damped second-order system. Values of the natural frequency

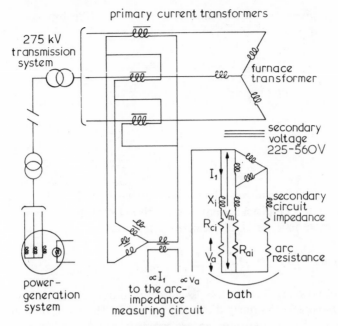

Fig. 3.4 *Electrical power supply system*

and damping coefficient have been estimated from plant trials using acceler-ometers connected to the electrode tip.

The single-phase furnace transmission system interconnecting the supply and arc furnace, illustrated in Fig. 3.4, can be represented by[2, 6]

$$I_1 = E(\alpha Z_{t2} - Z_{t3})\left(\sum_{i,\,l=1}^{3} Z_{ti} Z_{tl} \right) \qquad i \neq l \tag{3.11}$$

$$Z_{ti} = (R_{ai} + R_{ci}) + jX_i = R_{ti} + jX_i \tag{3.12}$$

where Z_{ti} are the total phase impedances referred to the transformer secon-dary and R_{ai} represent arc resistances, R_{ci} the system line resistances, X_i the line reactances, E the line voltage and α the complex 3-phase operator. The current magnitude may then be obtained in the form

$$I_1^2 = \frac{E^2\{(-R_{t3} - R_{t2}/2 - \sqrt{3}X_2/2)^2 + (\sqrt{3}R_{t2}/2 - X_2/2 - X_3)^2\}}{\left| \sum_{i,\,l=1}^{3} (R_{ti}R_{tl} - X_i X_l) \right|^2 + \left| \sum_{i,\,l=1}^{3} R_{ti} X_l \right|^2}$$

$$i \neq l \tag{3.13}$$

Linearizing by taking a first-order Taylor series expansion of Eqn. (3.13) with $I_1 = I_1^\circ + i_1$, $R_{a1} = R_{a1}^\circ + r_{a1}$ and assuming no interaction between the arc impedances gives the arc current/arc resistance relationship

i_1/r_{a1}

$$= \frac{(-I_1^\circ)^3 \left| (R_{t2} + R_{t3}) \sum_{i,\, l=1}^{3} (R_{ti} R_{tl} - X_i X_l) + (X_3 + X_2) \sum_{i,\, l=1}^{3} R_{ti} X_l \right|}{E^2 \{ (-R_{t3} - R_{t2}/2 - \sqrt{3}X_2/2)^2 + (\sqrt{3}R_{t2}/2 - X_2/2 - X_3)^2 \}}$$

$$i \neq l$$

or

$$\frac{i_1}{r_{ai}} = -F_1 \tag{3.14}$$

The arc discharge model relating arc voltage V_a and arc length H is based on Nottingham's empirical equation[7] which is an approximation to the static dc arc characteristic for similar electrode materials

$$V_{a1} = A_1 + D_1 H_1 + \frac{C_1 + B_1 H_1}{I_1^n} \tag{3.15}$$

In the electric-arc furnace the discharge is an ac arc and the electrodes are graphite and steel, obviously dissimilar materials. However, the inherent non-linearities of the high-current ac arc make its mathematical representation very difficult and experimental work on arcs of this power is almost nonexistent.[5]

For long high-power arcs Eqn. (3.15) can be approximated by the steady-state solution of Cassie's equation

$$V_{a1} = A_1 + D_1 H_1 \tag{3.16}$$

where D_1 is the arc discharge coefficient. The voltage signals V_m which provide a component of the error current in the amplidyne control field are measured at the transformer secondary terminals as indicated in Fig. 3.4. The voltage at this point is the resultant of the voltage across the distributed impedances in the supply cables and the resistance of the arc, and does not represent arc voltage. Arc voltage given by Eqn. (3.16) exists across the tip of the electrode and the scrap steel, and is almost impossible to measure on a production arc furnace. Consequently, the measured voltage magnitude may be expressed in the form[6]

$$V_{m1} = I_1 \{ (R_{a1} + R_{c1})^2 + X_1^2 \}^{1/2}$$

$$= \{ (A_1 + D_1 H_1 + I_1 R_{c1})^2 + I_1^2 X_1^2 \}^{1/2} \tag{3.17}$$

where I_1 represents arc current as a reference vector. Linearizing Eqn. (3.17) using a first-order Taylor series expansion and defining

$$|Z_{t1}| = \{ (R_{a1} + R_{c1})^2 + X_1^2 \}^{1/2}$$

$$I_1 = I_1^\circ + i_1 \quad \text{and} \quad R_{a1} = R_{a1}^\circ + r_{a1}$$

gives

$$v_{m1} = i_1 |Z_{t1}| + r_{a1} I_1^\circ (R_{a1}^\circ + R_{c1}) |Z_{t1}|^{-1}$$
$$= \{(D_1 h_1 + i_1 R_{c1})(R_{a1}^\circ + R_{c1}) + i_1 X_1^2\} |Z_{t1}|^{-1} \tag{3.18}$$

Including the relationship $i_1/r_{a1} = -F_1$ derived from the analysis of the transmission system gives the change in arc current as

$$i_1 = \frac{F_1 D_1 h_1}{F_1 R_{a1}^\circ - I_1^\circ} \quad \text{or} \quad i_1 = -W_1 D_1 h_1 = -A_1 h_1 \tag{3.19}$$

where the constant A_1 is usually called arc gain. Substituting Eqn. (3.19) into Eqn. (3.18) gives the relationship between changes in the measured voltage v_{m1} and arc length h_1

$$v_{m1} = D_1 h_1 \{(1 - W_1 R_{c1})(R_{a1}^\circ + R_{c1}) - W X_1^2\} |Z_{t1}|^{-1}$$

or

$$v_{m1} = D_1' h_1 \tag{3.20}$$

where D_1' is defined as the discharge coefficient.[6, 7] Eliminating v_{m1} and i_1 from Eqn. (3.18) and rearranging gives the arc resistance/arc length relationship[6, 7]

$$r_{a1} = \frac{h_1(D_1' + (WD)_1 |Z_{t1}|)}{I_1^\circ R_{t1} |Z_{t1}|^{-1}}$$

or

$$r_{a1} = B_1 h_1 \tag{3.21}$$

Equations (3.19) and (3.20) define the relationship between the change in arc length and the change in the current and voltage signals fedback to the arc-impedance measuring circuit

$$\epsilon_1 = -\{G_5 A_1 + G_4 D_1'\} h_1 \tag{3.22}$$

Small changes in the arc length have to be considered because only changes in electrode position can be conveniently measured on a production furnace.

Combining the models of the electrode position controller, arc discharge and transmission system provides a complete mathematical description of a single phase of the furnace control system as illustrated in Fig. 3.3. However, the derivation of this model involved making numerous assumptions which cannot be verified analytically. It was assumed that Nottingham's equation provides an adequate representation of the arc and that the interaction between the three regulators is negligible. Both these assumptions must be investigated before an improved control system can be designed using the derived model. This can best be achieved by conducting experiments on the furnace and identifying the properties of interest.[9]

3.4 Identification of the furnace control system

3.4.1 The arc discharge

The basic aims of the identification were threefold and included identification of properties of the arc discharge, investigation of the interaction between the regulators of an arc-impedance-controlled furnace and identification of a model of the Ward-Leonard drive or electrode position controller.

The diversity of the identification requirements entailed designing several experiments and recording a large amount of data. This was usually done in an iterative manner so that initial experiments added to the knowledge of the process and suggested the form of future experiments. In all the experiments the data were recorded on an analogue tape recorder prior to sampling and subsequent digital analysis.[6, 9, 10]

Initially, normal operating data for a typical melt were recorded to enable the relationship between electrode position, transformer secondary voltage and arc current to be investigated. A typical sequence of normal operating data is illustrated in Fig. 3.5.

The normal operating data were analysed to determine the discharge coefficient [Eqn. (3.20)] and the arc gain [Eqn. (3.19)]. During the first basket, at the beginning of the melt, the discharge coefficient D' was found typically to be 3764 V/m with an arc gain A of 859 kA/m. The values of the discharge coefficient (1653 V/m) and arc gain (367 kA/m) estimated during refining were notably lower than those experienced during the first basket.

Dynamic volt-ampere characteristics of the arc discharge in a production furnace during refining and at the beginning of the melt are illustrated in Fig. 3.6. The instability of the characteristics at the beginning of the melt

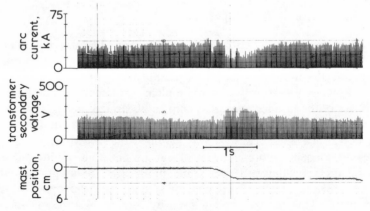

Fig. 3.5   *A typical sequence of normal operating data*

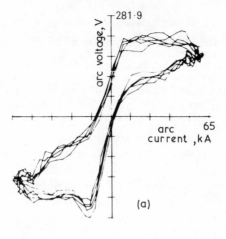

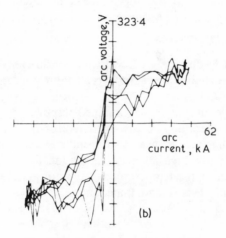

Fig. 3.6 *Dynamic arc characteristics*

indicates the dependence of the arc characteristics on the furnace environment.[6, 8, 9, 10]

These results tend to suggest that Nottingham's equation can only be used to represent the arc discharge when D'_1 and A_1 are functions of furnace environment. This can be substantiated by combining the experimental results with a simple analysis of the model.[8, 9]

Consider the effects of a step disturbance in the arc length over the period of a melt. When the furnace is cold, at the beginning of the melt, a step change in the arc length of amplitude h' will be sufficient to cause a current change i, thus

$$i = -A_{cold}h' \tag{3.23}$$

Towards the end of the melt, when the furnace is hot and the atmosphere ionized, a step change in arc length of amplitude h'' will be required to cause a similar change in arc current

$$i = -A_{hot}h'' \tag{3.24}$$

where $h'' > h'$, and hence $A_{cold} > A_{hot}$. Since the discharge coefficient D' is a function of the arc gain A and the resistance and reactance of the secondary conductors

$$D'_1 = \frac{\{(1 - W_1 R_{c1})(R_{a1}^\circ + R_{c1}) - W_1 X_1^2\}A_1}{W_1 |Z_{t1}|}$$

or

$$D'_1 = A_1 Q_1 \tag{3.25}$$

the relationship between the arc-impedance error function and the change in arc length [Eqn. (3.1)] can be expressed as

$$\epsilon_1 = -(G_5 + G_4 Q_1)A_1 h_1 \tag{3.26}$$

Hence ϵ_1, the error feedback signal, will vary considerably over the period of a melt. This will inevitably affect the overall loop gain and sensitivity of the electrode regulator and explains why the performance varies between the two extremes of unstable and highly overdamped responses.[8, 9]

A review of the literature on low-current arcs reveals that the electrical conductivity of the arc is heavily dependent upon arc temperature.[11] Moreover, at sufficiently high arc currents the localized ambient temperature can be shown to affect the temperature of the arc plasma which in turn will affect the relationship between arc impedance and arc length.[8] It would therefore appear that Eqns. (3.19) and (3.20) can only be used to represent high power ac arcs when D'_1 and A_1 are functions of ambient arc temperature.

3.4.2 Interaction between the regulators

Interaction between the three electrode regulators is minimized when operating under arc-impedance control. If the arc discharge had the properties of a metallic conductor when a disturbance occurred on one phase all the currents and voltages would change, but only the impedance of the disturbed phase would be altered. However, the resistance of low-power arcs of a given length depend on the arc current,[11] and if the current changes considerably the impedance may be affected. Unfortunately, no results are available for high-power ac arcs and hence these properties must be investigated experimentally.[6, 8, 9, 10]

Because there is no interaction between the three Ward-Leonard drives any interaction between the electrode regulators must be a combination of the mutual coupling of the high-current-carrying conductors, which can be shown to be less than 10%,[9] and the dependence of arc impedance upon arc current.

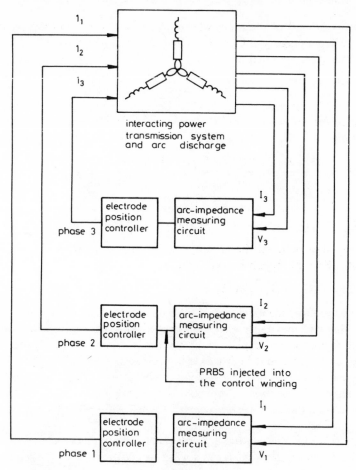

1_1

1_2

i_3

interacting power
transmission system
and arc discharge

phase 3

electrode
position
controller

arc-impedance
measuring
circuit

I_3

V_3

phase 2

electrode
position
controller

arc-impedance
measuring
circuit

I_2

V_2

PRBS injected into
the control winding

I_1

phase 1

electrode
position
controller

arc-impedance
measuring
circuit

V_1

Fig. 3.7 *A schematic diagram of the arc furnace control system*

Identification of the interaction between the electrode regulators is therefore a simple means of investigating the relationship between arc impedance and arc current for high-power ac arcs.

A schematic diagram of the three-phase system is illustrated in Fig. 3.7. A 127-bit 33·3-ms **PRBS** sequence was injected into the control winding of phase 2 electrode position controller, with the furnace in normal operation during refining. An on-line correlator was then used to identify the impulse responses relating the input on phase 2 to the three electrode positions. If there is any interaction between the regulators the cross-correlations will be representative of the impulse response of the interacting system. Inspection of the correlation functions (Fig. 3.8) confirms that there is minimal interaction when operating under arc-impedance control.

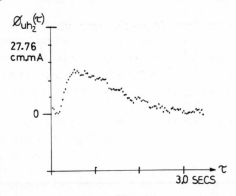

CCF. OF PHASE 2 CONTROL FIELD CURRENT
AND PHASE 2 MAST POSITION

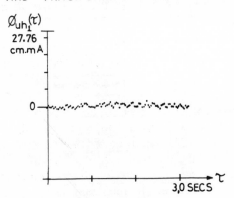

CCF. OF PHASE 2 CONTROL FIELD CURRENT
AND PHASE 1 MAST POSITION

Fig. 3.8 *Cross-correlation comparison*

However, the PRBS input may have been of insufficient power to excite the interacting modes of the system or overcome any dead zone or stiction associated with the regulators. Thus as a further check on the interaction, manual step disturbances were applied to the electrode position controller with the furnace in normal steady-state production during refining. The disturbances were applied by manually raising each electrode in turn and allowing the controller to re-establish the preset arc impedance with the other electrodes under normal automatic control. All the results indicated that the interaction is less than 10% confirming the correlation analysis.

The lack of interaction between the electrode regulators implies that arc impedance is practically independent of arc current. The regulators operating under arc-impedance control can therefore be considered as three independent electrode position controllers.[6, 12]

3.5 Identification of the electrode position controller

Having established that the three regulators can be considered as independent it is only necessary to identify a model of a single-phase electrode position controller. This was achieved by performing open-loop tests on the furnace with the arc discharge extinguished.[6, 10]

Initially, step inputs were applied to the open-loop system to check the linearity of the system over the operating range, estimate the system gain and assess the characteristics of the noise. Various PRBS sequences were then injected into the amplidyne control field and motor speed and electrode movement recorded. The signals were digitized and analyzed using an interactive identification package.[10] Many techniques including generalized least-squares parameter estimation, step and correlation analysis were applied to extract the maximum information from the data in an efficient manner, and a difference equation model was estimated

$$g_{kk}(z^{-1}) = \frac{z^{-2T}(0\cdot249z^{-1} + 0\cdot3079z^{-2} + 0\cdot095z^{-3}) \times 10^{-3}}{1 - 3\cdot547z^{-1} + 4\cdot826z^{-2} - 2\cdot9967z^{-3} + 0\cdot7177z^{-4}} \qquad (3.27)$$

$$k = 1, 2, 3, \qquad T = 1/24 \text{ s}$$

relating the arc-impedance error-signal to electrode position. Various model-order and validity checks were applied to ensure that the model provided an adequate representation of the electrode position controller.

Thus combining the identification results of the arc discharge, interaction between the regulators, and the electrode position controller provides a complete description of the furnace control system.

All the assumptions necessary in the derivation of the analytical model and the formulation of the identification results have been rigorously investigated on a production furnace. The analytical and identified models have been shown to have almost identical responses and the latter model can now be generalized to the three-phase case.

3.5.1 Three-phase models
Although arc-impedance control is widely adopted in arc furnace regulation, occasionally arc-current control is implemented. Three-phase models of arc-impedance- and arc-current-controlled regulators can now be derived based on the identification results for the electrode position controller, arc discharge and interaction between the phases.[12, 13]

3.5.2 Three-phase impedance-controlled model
The furnace transmission system illustrated in Fig. 3.4 can be represented by

$$I_1 = \frac{E(\alpha Z_{t2} - Z_{t3})}{\sum\limits_{k, l = 1}^{3} Z_{tk} Z_{tl}} \qquad k \neq l \qquad (3.28)$$

$$I_2 = \frac{E(Z_{t3} - \alpha^2 Z_{t1})}{\sum\limits_{k,\,l=1}^{3} Z_{tk} Z_{tl}} \qquad k \neq l \tag{3.29}$$

$$I_3 = \frac{E(\alpha^2 Z_{t1} - \alpha Z_{t2})}{\sum\limits_{k,\,l=1}^{3} Z_{tk} Z_{tl}} \qquad k \neq l \tag{3.30}$$

where the impedances are defined by Eqn. (3.12). The current magnitudes can then be evaluated as

$$I_1^2 = E^2 \frac{(-R_{t3} - R_{t2}/2 - \sqrt{3}X_2/2)^2 + (\sqrt{3}R_{t2}/2 - X_2/2 - X_3)^2}{Z_e} \tag{3.31}$$

$$I_2^2 = E^2 \frac{(R_{t3} + R_{t1}/2 - \sqrt{3}X_1/2)^2 + (X_3 + X_1/2 + \sqrt{3}R_{t1}/2)^2}{Z_e} \tag{3.32}$$

$$I_3^2 = E^2 \frac{(-R_{t1} + \sqrt{3}X_1 + R_{t2} + \sqrt{3}X_2)^2 + (-X_1 - \sqrt{3}R_{t1} + X_2 - \sqrt{3}R_{t2})^2}{Z_e} \tag{3.33}$$

where

$$Z_e = \left(\sum\limits_{k,\,l=1}^{3} (R_{tk}R_{tl} - X_k X_l) \right)^2 + \left(\sum\limits_{k,\,l=1}^{3} R_{tk} X_l \right)^2 \tag{3.34}$$

and $k \neq l$.

Linearizing using a first-order Taylor series expansion with $I_k = I_k^\circ + i_k$, $k = 1, 2, 3$ and $R_{ai} = R_{ai}^\circ + r_{ai}$ in Eqns. (3.31) to (3.33) for $i = 1$, $i = 2$, $i = 3$ respectively, and assuming zero interaction between the three arc resistances and an infinitely stiff supply voltage yields, analogous to Eqn. (3.14),

$$i_k = -F_k r_{ak} \qquad k = 1, 2, 3 \tag{3.35}$$

Similarly extending the derivation of the arc model in Eqns. (3.15) to (3.21) to three phases and assuming zero interaction yields

$$\epsilon_k = -\{G_5 A_k + G_4 D_k'\} h_k \qquad k = 1, 2, 3 \tag{3.36}$$

which combined with the identified model of the electrode position controller

$$h_k = g_{kk}(z^{-1}) \epsilon_k \qquad k = 1, 2, 3 \tag{3.37}$$

where $g_{kk}(z^{-1})$ as given by Eqn. (3.27), defines the three-phase arc-impedance-controlled furnace model.[12, 13]

3.5.3 Three-phase current-controlled model

In a three-phase arc furnace operating under current control, the electrode position controllers attempt to maintain the arc current at their reference values. If a disturbance occurs, all the arc currents will change and the three electrode regulators operate to re-establish the desired power input level. The regulation is therefore spread over all three electrode regulators which are no longer non-interacting.

Although it is assumed that there is no inherent interaction between the arc impedances, the operation of all the regulators to clear a disturbance suggests that the arc currents can be represented by a relationship of the form

$$i_k = f_k(r_{a1}, r_{a2}, r_{a3}) \qquad k = 1, 2, 3 \tag{3.38}$$

Such a relationship can be obtained by linearizing the current magnitude, in Eqns. (3.31) to (3.33), using a first-order Taylor series expansion with $I_k = I_k^\circ + i_k$, $R_{ak} = R_{ak}^\circ + r_{ak}$, $k = 1, 2, 3$ to yield

$$i_1 = E^2 \frac{\alpha_1 r_{a1} + \beta_1 r_{a2} + \gamma_1 r_{a3}}{I_1^\circ Z_e^2} \qquad k \neq l \tag{3.39}$$

$$i_2 = E^2 \frac{\alpha_2 r_{a1} + \beta_2 r_{a2} + \gamma_2 r_{a3}}{I_2^\circ Z_e^2} \qquad k \neq l \tag{3.40}$$

$$i_3 = E^2 \frac{\alpha_3 r_{a1} + \beta_3 r_{a2} + \gamma_3 r_{a3}}{I_3^\circ Z_e^2} \qquad k \neq l \tag{3.41}$$

relating the change in the arc currents to the changes in the arc resistances as the latter are adjusted by the electrode regulators. Eliminating r_{a1}, r_{a2} and r_{a3} using Eqn. (3.21) gives the current-controlled transmission system model

$$\begin{pmatrix} i_1 \\ i_2 \\ i_3 \end{pmatrix} = E^2/Z_e^2 \begin{pmatrix} \dfrac{B_1\alpha_1}{I_1^\circ} & \dfrac{B_2\beta_1}{I_1^\circ} & \dfrac{B_3\gamma_1}{I_1^\circ} \\[2mm] \dfrac{B_1\alpha_2}{I_2^\circ} & \dfrac{B_2\beta_2}{I_2^\circ} & \dfrac{B_3\gamma_2}{I_2^\circ} \\[2mm] \dfrac{B_1\alpha_3}{I_3^\circ} & \dfrac{B_2\beta_3}{I_3^\circ} & \dfrac{B_3\gamma_3}{I_3^\circ} \end{pmatrix} \begin{pmatrix} h_1 \\ h_2 \\ h_3 \end{pmatrix} \tag{3.42}$$

with $k \neq l$ and where α_k, β_k and γ_k are constants which depend upon the arc characteristics and the impedance of the secondary conductors.

Combining Eqn. (3.37) with Eqn. (3.42) and the current-controlled model for the error feedback to the amplidyne

$$\epsilon_k = -N_k i_k \qquad k = 1, 2, 3 \tag{3.43}$$

where N_k is a constant feedback gain, defines the interacting three-phase current-controlled furnace model[9, 12] illustrated in Fig. 3.9.

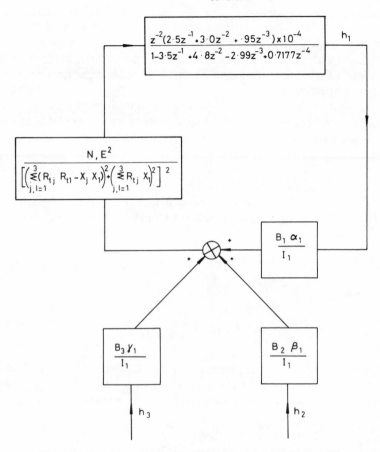

electrode position
controller

$$z^{-2}(2 \cdot 5z^{-1} + 3 \cdot 0z^{-2} + \cdot 95z^{-3}) \times 10^{-4}$$

over

$$1 - 3 \cdot 5z^{-1} + 4 \cdot 8z^{-2} - 2 \cdot 99z^{-3} + 0 \cdot 7177z^{-4}$$

h_1

$$\frac{N, E^2}{\left[\left(\sum_{j,l=1}^{3}(R_{lj} R_{t1} - X_j X_1)\right)^2 + \left(\sum_{j,l=1}^{3} R_{lj} X_1\right)^2\right]^2}$$

$$\frac{B_1 \, \alpha_1}{I_1}$$

$$\frac{B_3 \, \gamma_1}{I_1}$$

$$\frac{B_2 \, \beta_1}{I_1}$$

h_3

h_2

Fig. 3.9   *A schematic diagram of the currrent controlled model*

Note that the arc characteristics D'_k, A_k and B_k in the three-phase models depend upon the ambient arc temperature and can therefore take a range of values as identified in previous sections. This also implies that the interaction between the regulators of a current-controlled furnace [Eqn. (3.42)] will depend upon the characteristics of the arc of each phase. Localized changes in the arc environment, which may be caused by vaporization of impurities in the steel, will therefore affect the degree of interaction between the phases.

The current- and impedance-controlled models have been formulated assuming equal arc characteristics and electrode-position-controller dynamics. Such conditions often prevail in arc furnace operation. However, the models are not restrictive and can be used to simulate the effects of unequal arc characteristics or electrode-position-controller dynamics if required.

3.6 Conclusions

Three-phase models of an arc-impedance- and an arc-current-controlled electric-arc furnace have been derived by combining analytically derived models with the results of an identification study. Although the simplifying assumptions which are often necessary in the derivation of an analytical model can usually be made with confidence, based on the available literature and experience of modelling similar processes, this approach was precluded in the case of the electric-arc furnace because of the lack of detailed knowledge of the high-power arc discharge. However, by deriving a simple single-phase model to gain insight into the process operation, experiments were designed and the modelling assumptions were investigated using identification techniques. It was shown that properties of the electric-arc discharge are heavily dependent upon ambient arc temperature and that arc impedance is virtually independent of arc current in arcs of this power. By incorporating this information in the single-phase model and extending this to a three-phase representation using the identified model of the electrode position controller, an accurate mathematical description of an arc-impedance- and arc-current-controlled furnace was derived.

The three-phase models have been used successfully to design improved control schemes for the arc furnace[13] including a dual impedance/current control strategy,[12] a temperature weighting adaptive controller[8] and p.i.d. and minimum variance regulators.[9]

3.7 References

1 BROWN, P., and LANGMAN, R. D.: 'Simulation of closed loop energy control applied to arc furnaces', *J. Iron Steel Inst.*, August, 1967, pp. 837–847
2 NICHOLSON, H., and ROEBUCK, R.: 'Simulation and control of electrode position controllers for electric arc furnaces', *Automatica*, **8**, 1972, pp. 683–693
3 MORRIS, A. S., and STERLING, M. J. H.: 'Analysis of electrode position controllers for electric arc steelmaking furnaces', *Iron Steel Internat.*, **48**, August, 1975, pp. 291–298
4 BOWMAN, B., JORDAN, G. R., and FITZGERALD, F.: 'The physics of high current arcs', *J. Iron Steel Inst.*, 1969, pp. 798–805
5 BOWMAN, B., JORDAN, G. R., and WAKELAM, D.: 'Electrical and photographic measurements of high power arcs', *J. Phys. D. Appl. Phys.*, 3, 1970, pp. 1089–1099
6 BILLINGS, S. A., and NICHOLSON, H.: 'Identification of an electric arc furnace electrode control system', *Proc. IEE*, **122**, 1975, pp. 849–856
7 BROWNE, T. E. (Jnr.): 'The electric arc as a circuit element', *J. Electrochem. Soc.*, January, 1955, pp. 27–37
8 BILLINGS, S. A., and NICHOLSON, H.: 'Temperature weighting adaptive controller for electric arc furnaces', *Ironmaking and Steelmaking*, 4, 1977, pp. 216–221
9 BILLINGS, S. A.: 'Modelling, identification and control of an electric arc furnace', Ph.D. thesis, Sheffield University, 1975

10 BILLINGS, S. A., STERLING, M. J. H., and BATEY, D. J.: 'SPAID—an interactive data analysis package and its application to the identification of an electric arc furnace control system', *IEE Conf. Publ. 159*, Random Signals Analysis, April, 1977, pp. 161–170

11 HAYDON, S. C. (Editor): 'Discharge and plasma physics', (Armidale, NSW, University of New England, 1964)

12 BILLINGS, S. A., and NICHOLSON, H.: 'Modelling a three-phase electric arc furnace; a comparative study of control strategies', *Appl. Math. Modelling*, 1, 1977, pp. 355–361

13 BILLINGS, S. A., BOLAND, F. M., and NICHOLSON, H.: 'Electric arc furnace modelling and control', *Automatica*, 15, 1979, pp. 137–149

Chapter 4

Development of models for shape-control-system design

M. J. Grimble

List of principal symbols

v_1 strip input velocity, m/s
v_2 strip exit velocity, m/s
h_1 strip input gauge (with tension)
h_2 strip output gauge (with tension)
h_{10} strip input gauge (with no applied tension)
h_{20} strip output gauge (with no applied tension)
w strip width, m
μ coefficient of friction in the roll gap
F roll force, N
δ reduction $(h_1 - h_2)$
c constant in Hitchcock's formula
σ_0 mean output stress, N/m^2
$\sigma(x)$ stress at point x across the output strip width
E Young's modulus of elasticity, N/m^2
I second moment of area, m^4
k elastic foundation constant, N/m
$p(x)$ specific roll force, N/m
v Poisson's ratio
$\bar{h}_2$ mean output gauge

4.1 Introduction to the shape-control problem

4.1.1 General introduction

In recent years the problem of the control of the gauge of steel strip, exiting from a rolling mill, has largely been solved.[1] The major problem of current interest in cold-rolling mills involves the control of internal stresses in the rolled strip.[2] This is referred to as shape control which is an unfortunate misnomer that often causes confusion. Strip with good shape does not have internal stresses rolled into it. When such strip is cut into sections the strip remains flat when laid on a flat surface. Whilst the strip is being rolled it is under very high tensions and shape defects are often not apparent to the eye (this is referred to as latent shape). It might therefore be expected that shape measurement is a difficult problem. Reliable shape-measuring devices have only become available in the last ten years and this has enabled recent work on shape control to progress to the closed-loop control stage.[3]

To illustrate how bad shape might occur consider the situation where the entry gauge profile to a mill stand is of uniform thickness. Assume that the work rolls in the stand are deformed so that the strip existing from the stand is thicker in the central region than at the edges of the strip. In the absence of lateral spread, any differential reduction tends to produce differential elongations in adjacent longitudinal elements of the strip.[4] That is, due to the mass-flow relationships the strip tends to be longer at the edges than in the central regions. Since the strip is one homogeneous mass such differential elongations cannot occur and internal stresses result. If these stresses are sufficiently large the long edges to the strip will appear as wavy edges which can be observed whilst rolling is in progress (this is called manifest shape and is illustrated in Fig. 4.1).

Shape- and gauge-control systems interact and they should be designed together to give the best overall performance. A gauge-corrective action, by altering the forces in the roll gap, influences the shape. Similarly, a shape-correcting action normally alters the forces acting in the rolling stand which affects the output gauge profile. Ideally shape- and gauge-control systems should be combined into one integrated scheme. However, since shape-

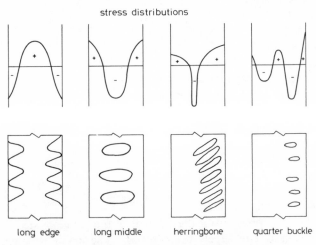

Fig. 4.1 *Manifest buckling forms*

control systems are often added to existing steel mills this is not always possible. In the present discussion the effect of the gauge-control loop will be neglected, except for the disturbance it introduces into the shape-control scheme. The instability which can arise due to shape- and gauge-control-system interaction, will therefore be neglected.

The first requirement in the design of a shape-control-system for a cold-rolling mill is for a model of the mill. Both static and dynamic mill models are required. The static model enables shape profiles to be calculated for a given set of actuator positions and is used to generate the steady-state mill gains which are used in the dynamic model. The dynamic model is a simulation of the state equations for the complete system, including the shapemeter and strip dynamics. In the following, the static and dynamic models for a Sendzimir cold-rolling mill are described. A Sendzimir mill is mechanically complex as may be judged from the roll configuration in Fig. 4.2. There has been very little previous published work on either model development or shape control for such mills.[5, 6]

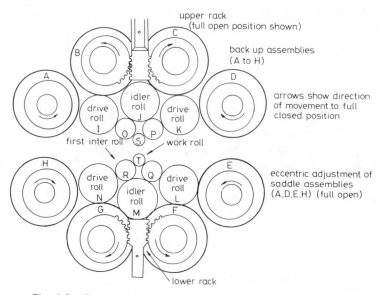

Fig. 4.2 *Roll and back-up bearing arrangement and adjustment*

4.1.2 Definition of strip shape

Before considering a formal definition of strip shape the stresses acting in the strip will be discussed.[7] The transverse variation in the longitudinal stresses is caused by:

1. Any transverse variations in the reduction in the stand.
2. Transverse variations in the slip and hence the strip velocity at the exit of the stand (or at the entry to the next stand in a multistand mill).

The (output) slip is defined as

$$f = \frac{\text{strip output velocity} - \text{peripheral velocity at } \phi_n}{\text{peripheral velocity at } \phi_n} \quad (4.1)$$

but from mass flow $v_2 h_2 = v_n h_n$ and from geometry the peripheral velocity $v_n = $ (strip horizontal velocity at ϕ_n)/cos ϕ_n. Thus $v_n = v_2 h_2/(h_n \cos \phi_n)$ and the slip becomes

$$f = \frac{h_n \cos \phi_n - h_2}{h_2} \simeq \frac{h_n - h_2}{h_2} \quad (4.2)$$

A transverse variation in the reduction will clearly cause a differential elongation in adjacent strip segments and will therefore affect the strip shape. The effect of slip variations is not so obvious. However, from the above relationship

$$\text{Output velocity } v_2 = (1 + f)v_n \quad (4.3)$$

and thus an increase in the neutral angle from ϕ_{n_1} to ϕ_{n_2} will increase the slip and hence output velocity (note that at the neutral angle $v_{n_1} = v_{n_2}$). It follows that if the slip varies across the strip width, then the output velocity will also vary.

4.1.3 Units of shape—mon

Pearson[2] has defined a unit of shape called the 'mon' in terms of the classical long-edge or long-middle defect. Pearson relates shape to the amount of bowing present in narrow bands slit from the strip. The 'mon' defines the shape of strip which if slit into bands 1 cm wide, would produce a lateral curvature corresponding to a radius of 10^4 cm. If this definition is applied to a long-edge or long-middle defect, the shape in mons is the fractional difference in elongation between the centre and edge of the strip multiplied by a factor of 10^4. That is, let Δl represent the difference in length between the longest and shortest line segments of the strip, then

$$\text{Mon} = \frac{\Delta l}{l} 10^4$$

For example, for 0.01% elongation, $\Delta l/l = 0.0001$ and this equals one mon unit. The strip shape may be defined as the relative length difference per unit width expressed in mon cm^{-1}. Hence

$$\text{Shape in mon cm}^{-1} = \frac{\Delta l}{l} \frac{10^4}{w} \quad (4.4)$$

where w is the strip width in cm. A specific shape of 0.2 mon/cm is representative of the best shape obtainable by direct rolling without subsequent flattening treatment.

4.1.4 Units of shape—I units

Sivilotti[8] et al. express the quantity $\Delta l/l$ in I units where

$$\text{I units} = \frac{\Delta l}{l} 10^5 \tag{4.5}$$

4.2 Basic components in the Sendzimir mill system

4.2.1 Introduction

The basic components in the Sendzimir mill shape-control system are illustrated in Fig. 4.3. The construction of the mill is described in the following section and the shape measurement system is then discussed. The models for these systems are then considered, beginning in Section 4.3.

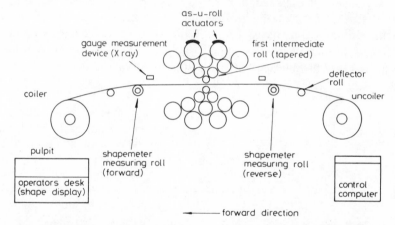

Fig. 4.3 *Schematic diagram showing the basic components in the system*

4.2.2 The Sendzimir mill

There are various types of Sendzimir mills but the mill considered here is a cluster mill as shown in Fig. 4.2. The mill considered is 1·6 metres wide and is used for rolling stainless steel. The motor drive is applied to the two outer second intermediate rolls (I, K, L and N in Fig. 4.2) and the transmission of the drive to the work rolls is due purely to interroll friction. The rolls labelled I to T have thrust bearings and are thus free to float. The outer rolls (A–H) are as shown in Fig. 4.2. The outer rolls are supported by eight saddles per shaft which are fixed to the mill housing. These saddles contain eccentric rings which are free to rotate in the circular saddle bores. These assemblies are used in the screwdown and shape-control mechanisms.

The screwdown racks act upon assemblies B and C and F and G. Only the top assemblies (B and C) have both screwdown and shape-actuator eccentric rings (these actuators are referred to as As-U-Rolls by the manufacturers of Sendzimir mills). The screwdown eccentric ring is keyed to the appropriate back-up roll shaft. Thus a rack position change causes the screwdown ring and the shaft to rotate. This allows the shaft centre to be moved towards or away from the mill housing. The shaft is, of course, keyed to the screwdown eccentric rings in all eight saddles. Essentially the screwdown system enables the average load to be varied during rolling without roll bending.

The As-U-Roll eccentric rings on shafts B and C enable roll bending to be achieved, for shape control, during rolling. Each of the saddles supporting these two shafts is fitted with an extra eccentric ring, situated between the saddle and the screwdown eccentric ring. This eccentric ring may be moved independently of the shaft and screwdown eccentric ring. There are eight As-U-Roll racks which are capable of individual adjustment to vary the positions of these eccentric rings. Thus a different displacement between the shafts and the housing at each saddle position may be achieved. When the screwdown system is operated, the bearing shafts and screwdown eccentrics rotate at each saddle simultaneously. The As-U-Roll eccentrics do not rotate. Thus, the bending profile set up by the As-U-Roll is maintained, the shafts merely moving to a new position parallel with the original.

There is also an indirect form of shape control device in the form of the first intermediate rolls (O, P, Q and R). The top rolls O and P are tapered at the front of the mill and the bottom rolls Q and R are tapered at the back of the mill. These rolls may be moved in pairs axially in and out of the cluster and thus the pressure at the edges of the strip may be controlled within certain limits. For example, by moving the bottom rolls out of the front of the mill, the pressure at the leading edge of the strip may be reduced.

The roll assemblies A, D, E and H are collectively termed the 'side eccentrics'. The rolls A and H can be adjusted independently from D and E and thus the mill need not be symmetrical about a vertical centre line. The positioning of the side eccentrics can affect the steady-state mill gains discussed in Section 4.3.7. These eccentrics are used for the fine positioning of the pass line (the strip centre line) and are normally set at the commencement of a rolling pass and not changed during the pass. The left-hand eccentrics control the positions of rolls A, H, I, N, O, R, S and T and the right-hand eccentrics control the remainder.

For modelling purposes it is worth noting that although the As-U-Roll eccentrics and screwdown eccentrics have the same common shaft the shape and screwdown systems are essentially non-interactive. The shape-control system proposed involves continuous use of the As-U-Rolls but only intermittent use of the first intermediate rolls. This two-level strategy simplifies the design stage.

4.2.3 The shape-measurement subsystem
There are various types of shape-measuring devices commercially available.[13, 14] The most successful and reliable devices seem to be the Loewy

Robertson Vidimon shapemeter[9] and the ASEA Stressometer shapemeter.[10, 11] The Japanese have already applied the former to open-loop control on a Z mill.[12] However, the latter device will be considered here since this is the instrument employed on the steel mills of interest.

The stressometer measuring roll is divided into a number of measuring zones (in this case 31) across the roll. The stress in each zone is measured independently of that in adjacent zones. The sensors are a form of magnetoelastic force transducer and are placed in four slots, equally spaced around the roll periphery. The periodic signals from each zone are filtered and the stress in each zone $\sigma(x)$ is calculated. The average stress σ_0 is also calculated and the difference $\Delta\sigma(x) = \sigma(x) - \sigma_0$ is displayed (or used for feedback purposes) on a separate instrument for each measuring zone.

In developing a model for the shape-measurement subsystem there are various factors to be considered. For example, the shapemeter filter, referred to above, is changed with line speed and thus the dominant time-constant may have one of five different values (in the range 4·35 to 0·11 seconds). From observations these smoothed signals seem to contain a white-noise component. The number of measuring zones which are covered by the strip depends upon the strip width and thus not all zones are active normally.

4.3 Static model for a Sendzimir mill

4.3.1 Introduction to the static model
The study of any scheme for control of strip shape must be preceded by an accurate analysis of the formation of the loaded roll gap in the rolling stand. A static model for the Sendzimir mill is therefore required which will include all mechanical force-deformation relationships. It is important for control purposes to note that these relationships are both non-linear and schedule (or mill-product) dependent.

The static model must allow for the bending and flattening of the rolls in the mill cluster and for the plastic deformation of the strip in the roll gap. The model should provide:

(*a*) Mill gains between shape actuator movements and strip shape changes, based upon a small change calculation. These represent a linearized set of mill gains for a given operating point.
(*b*) Details of the degree of control which may be achieved with a given shape actuator.
(*c*) Insight into the mechanisms involved in the roll cluster and the roll gap which affect strip shape.

A static model was developed[5, 6] in the form of a Fortran IV computer program. The model enables the output shape profile to be calculated corresponding to a given set of shape-actuator (As-U-Roll) rack positions. The model also enables the shape change due to a change in the roll cambers to be calculated. Such a change can result from movement of the first intermediate rolls or from a temperature change in the work rolls.

The static model involves four main sets of calculations which may be listed as follows:

(a) Roll bending calculations, based upon simple beam theory, for calculating roll bending due to a given distributed load.
(b) Roll flattening calculations which enable the roll flattening between two rolls to be found for a given load.
(c) Roll force calculations which enable the roll force and pressure to be calculated for given strip dimensions and properties.
(d) Output gauge- and shape-profile calculations which enable these profiles to be determined corresponding to given interroll pressure and deflection profiles.

The assumptions made in deriving the static model described here may be listed as:

(i) Elastic recovery of the strip may be neglected.
(ii) Horizontal deflections of the rolls may be neglected.
(iii) The centre-line strip thickness is assumed to be specified.
(iv) Plane strain conditions exist in the roll gap.
(v) The mill is symmetrical about a line passing through the work roll centres (equal settings on the side eccentrics).
(vi) Strip edge effects may be neglected.
(vii) Deflections due to shear stresses may be neglected.

The validity of the first assumption rests upon the small work rolls used in Z-mills which limit the arc of contact and give sharp roll-gap angles. The second assumption that there is no appreciable deflection in the rolling direction follows from the lateral support given to the work rolls by the roll cluster. Because shape control depends upon the profile of the loaded roll gap, it is assumed that the stand is operating under automatic gauge control and assumption three follows.

4.3.2 Roll-bending calculations
All the rolls in the middle of the mill cluster are resting upon one another and thus each roll may be considered to be resting upon an elastic foundation. The deflection of a beam resting upon an elastic foundation due to a point force may be calculated by assuming that the deflection is proportional to the reaction at that point. The bending deflection y can therefore be calculated as a function of the applied force F and the distance x from one end of the beam, that is $y = f(F, x)$. To be more specific, if l is the length of the roll and the force F is applied at a point $x = a$, then[6]

$$y(x) = \frac{(F + k\,\Delta y(x))}{k} \cdot \lambda \cdot \beta(\lambda, l, a) \qquad (0 \le x \le a) \tag{4.6}$$

where λ is a constant for a given roll

$$\lambda \triangleq \left(\frac{k}{4EI}\right)^{1/4} \tag{4.7}$$

and $\beta(\lambda, l, a)$ is a function of λ and the length l and a. The gap between the unloaded roll and the foundation is denoted by $\Delta y(x)$. The above equation is derived from the solution to the differential equation

$$EI\frac{d^4y}{dx^4} = F - k(y - \Delta y) \tag{4.8}$$

which follows from simple roll-bending theory. If the point x is such that $l \geq x > a$ the deflection may be calculated using the above result but with a replaced by $(l - a)$ and with x replaced by $(l - x)$.

4.3.3 Roll-flattening calculations

The calculation of the deformation which occurs between two touching rolls in the cluster or between the work rolls and the strip is discussed below. The approach follows that of Timoshenko and Goodier[15] and more recently that of Edwards and Spooner.[7] The roll surfaces may be assumed to be cylindrical, neglecting minor bending distortions. Now when two infinitely long elastic cylinders are in contact the total interference y' can be written as a function of the load per unit length q'

$$y'_{12} = q'(c_1 + c_2) \log_e\left(\frac{(e)^{2/3}(d'_1 + d'_2)}{2q'(c_1 + c_2)}\right) \tag{4.9}$$

where d'_1 and d'_2 are the diameters of the cylinders and c_1 and c_2 are two elastic constants. The loading along a roll is of course non-uniform and the roll is also of finite length. However, the influence of a point load does not extend far along the roll and neglecting second-order errors, q' may be replaced by the interroll specific force $q(x)$, to calculate the interference $y_{12}(x)$.

Let F denote the total load on the rolls; then since $q(x) \simeq F/w$,

$$\log_e(q(x)w/F) \ll \log_e\left(\frac{(e)^{2/3}(d_1 + d_2)w}{2F(c_1 + c_2)}\right)$$

Using this approximation

$$q(x)$$

$$= \frac{y_{12}(x)}{(c_1 + c_2)(\log_e((e)^{2/3}/2(c_1 + c_2)) + \log_e(d_1 + d_2) - \log_e(F/w))} \tag{4.10}$$

If $y_{12}(x) < 0$ the specific force $q(x)$ must be set to zero.

An equation is now required which will enable the interference to be calculated corresponding to a given force distribution $q(x)$. Let $y_1(x)$ and $y_2(x)$ denote the deflections of rolls 1 and 2 calculated from the roll-bending equations and let D_{12} denote the distance between the roll centres. Also note that the interference depends upon the thermal and ground camber $y_c(x)$. The interference between the two rolls is thus

$$y_{12}(x) = \frac{d_1 + d_2}{2} + 2y_c(x) - D_{12} + y_2(x) - y_1(x) \tag{4.11}$$

Equations (4.10) and (4.11) may be solved iteratively to calculate the specific force and the interference. The algorithm is judged to have converged when the following force balance equation is satisfied to a given accuracy:

$$\int_0^w q(x)\, dx = F \tag{4.12}$$

Orowan has previously noted that extensive work-roll flattening can occur.[16] The transverse profile of the roll gap is significantly affected by the indentation of the roll by the strip across the strip width. This situation can be modelled by the use of work-roll influence functions and convolution integrals. However, for the present discussion approximate results are used based upon the work of Edwards and Spooner.[7] They noted that the roll flattening was slightly dependent upon specific roll force and related to the Hertzian flattening which occurs between two elastic cylinders of the same diameter d. Thus the work-roll flattening is given by

$$y_{ws}(x) = (b_1 + b_2 p(x)) y_h(x) \tag{4.13}$$

where

$$y_h(x) = 2p(x)c \frac{\log_e(e)^{2/3}\, d}{(2p(x)c)} \tag{4.14}$$

This latter equation follows from Eq. (4.9) by setting $c_1 = c_2 = c$ and $d_1 = d_2 = d$. The constants b_1 and b_2 are estimated using plant test results and $c \triangleq (1 - v^2)/(\pi E)$. Equations (4.13) and (4.14) must also be solved iteratively.

4.3.4 Roll-force calculations

The roll-force calculations are an important part of the static model.[17, 18] The conditions within the roll gap are represented in the roll-force model. The equations governing the plastic reduction of the strip in the roll gap are contained in this model. Both explicit and implicit methods of roll-force calculations have been used in the static model. For present purposes the explicit calculation procedure only is described. The following formula was developed by Bryant and Osborn:[19]

$$F = F_0/(1 - bF_0 - 0.4aB°) \tag{4.15}$$

where

$$a = 1.4 \sqrt{\frac{h_2}{h_1}} \left(\frac{\mu}{h}\right)^2 Rc \tag{4.16}$$

$$B° = (\bar{k} - \bar{\sigma})\sqrt{R\delta} \tag{4.17}$$

$$b = \left(\frac{c}{2\delta} - \frac{F°}{2}\left(\frac{c}{2\delta}\right)^2\right) \tag{4.18}$$

and $F°$ is the roll force obtained when the roll is undeformed and of nominal radius R. Also note that $\bar{k} \triangleq (k_1 + 2k_2)/3$, $\bar{\sigma} \triangleq (2\sigma_1 + \sigma_2)/3$ and $\bar{h} \triangleq 0.72h_2 + 0.28h_1$.

The calculation time makes it necessary to use an explicit rather than an implicit roll-force calculation procedure. For modelling purposes the width of the strip is split into 25-mm sections and the roll force must be calculated in each of these sections. Thus for one-metre-wide strip there are 40 strip sections. The shape calculation is iterative and thus all roll-force calculations must be performed for each iteration of the shape algorithm. Thus, the total roll-force calculation time is multiplied by the number of strip sections and the number of shape-program iterations. Iterative roll-force algorithms are therefore not used in the present mill model although one algorithm, employing Hitcock's formula, was used for comparison with the explicit model.[20] The calculated roll forces differed by less than 1% in many cases.

4.3.5 Output-gauge- and shape-profile calculations

The equations governing the output gauge and shape profiles are obtained below. The output-gauge profile is determined by the combined effects of roll bending, thermal and ground cambers and differential strip flattening. The change in the gauge profile due to these effects is given by

$$\Delta h_2'(x) = 2(y_{ws}(x) - \bar{y}_{ws}(x)) + y_u(x) + y_l(x) + 2y_c(x) \tag{4.19}$$

where y_{ws} and $\bar{y}_{ws}$ represent the interference and mean interference respectively, between the work roll and the strip, y_u and y_l represent the upper and lower work-roll deflections respectively, and y_c represents the total work-roll camber. The mean deviation in the output gauge is given by

$$\Delta h_2(x) = \frac{1}{w}\int_0^w \Delta h_2'(x)\,dx \tag{4.20}$$

and then the deviation from the mean is given by

$$\Delta h_2(x) = \Delta h_2'(x) - \Delta\bar{h}_2 \tag{4.21}$$

The new output gauge deviation is calculated from the iterative equation

$$\Delta h_2^{k+1}(x) = \Delta h_2^k(x) - \alpha(\Delta h_2^k(x) - \Delta h_2'(x)) \tag{4.22}$$

where the convergence scalar α is chosen to give a stable solution. The new gauge profile is calculated using the following result:

$$h_2(x) = \bar{h}_2 + \Delta h_2^{k+1}(x) \tag{4.23}$$

where $\bar{h}_2$ is the mean output gauge.

The new input and output stresses can be calculated using the new gauge profile and the following results due to Edwards and Spooner:[7]

$$\Delta\sigma_2(x) = \beta E\left(\frac{h_2(x)}{h_1(x)} \cdot \frac{\bar{h}_1}{\bar{h}_2} - 1\right) + \frac{\Delta\sigma_0(x)}{(1 + \gamma)} \tag{4.24}$$

$$\Delta\sigma_1(x) = \gamma \, \Delta\sigma_2(x) \tag{4.25}$$

where

$$\gamma \triangleq \frac{\sigma_1(x) - \bar{\sigma}_1}{\sigma_2(x) - \bar{\sigma}_2} \tag{4.26}$$

β is a constant ($\beta \simeq 0.5$), and E denotes Young's modulus of elasticity.

The work-roll indentation is particularly important at the edges of the strip. The most accurate way of modelling this situation is by the use of work-roll influence functions. The roll profile change may then be calculated from the convolution of the pressure profile and the influence function. However, for the present discussion the roll flattening results described in Section 4.3.3 will be employed.

4.3.6 Static-model computer algorithm

The static-model computer algorithm enables a change in the shape profile due to a change in the rank position, and thence the gains of the mill, to be calculated. The flow chart for the main program is shown in Fig. 4.4.

The program begins by initializing all the variables and the roll force is then calculated using the roll-gap model. Now symmetry about a line passing through the work-roll centres can be assumed so that calculations are necessary only for the left side of the cluster. The subroutine BEND calculates the pressure profiles and roll profiles of one half of either the top or bottom cluster. If symmetry is not assumed then the routine BEND has to be called four times to calculate all the pressure and roll profiles. At the end of each iteration a convergence test is carried out on the shape profile. The above calculations are repeated until the error between two successive shape profiles is less than a predetermined value.

The pressure and roll profile calculation procedure is illustrated in Fig. 4.5. Only the top half cluster is shown in this figure and the thick lines are drawn to show the path of calculation. The small circles, denoted c_1, c_2, c_3, c_4,

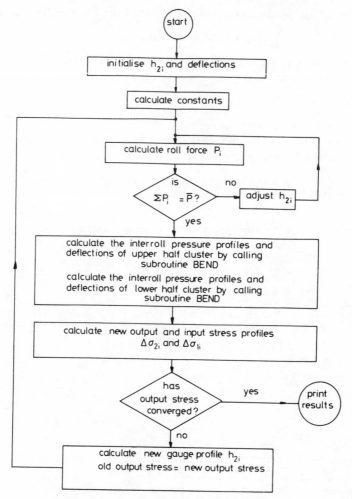

Fig. 4.4 *Flow chart for the main program*

represent an iterative process for one particular interroll pressure calculation. A satisfactory convergence in pressure is shown by the letter Y and non-convergence is represented by N. As the pressure profile between two rolls depends on the interference or flattening of the rolls and the interference is itself dependent on the pressure, the process is iterative.

Figure 4.6 shows the flow chart for an iterative calculation of interroll pressure. Here y_1 and y_2 are the deflections of two rolls in contact and $q(n)$ is the specific pressure profile between them.

4.3.7 Static-model results
The static model provides two main types of results for a given rolling schedule. The first category is concerned with the physical processes and

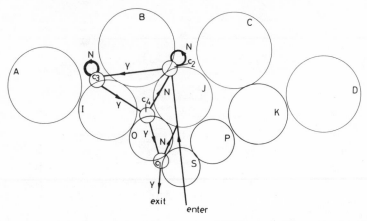

Fig. 4.5   *Path of calculation for subroutine BEND to calculate pressure profiles*

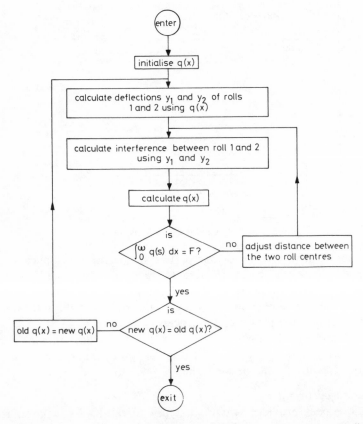

Fig. 4.6   *Flow chart for calculating the interroll pressure between two rolls*

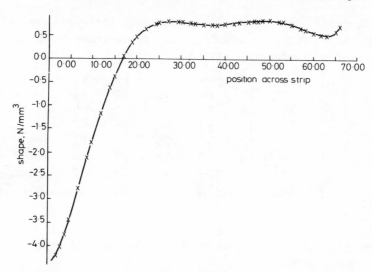

Fig. 4.7 *Shape change due to 1-mm movement in rack 1 (for 1.7-m wide strip and rolls without camber)*

includes the calculated shape, gauge and pressure profiles. These are important when the degree of control of a particular profile is of interest. The shape profiles corresponding to representative As-U-Roll rack positions are shown in Figs. 4.7 and 4.8. The greatest shape change is of course under the particular shape actuator which has been varied from the null position.

The mill gains represent the second major source of information available from the static model. Linearized mill gains are calculated about a given

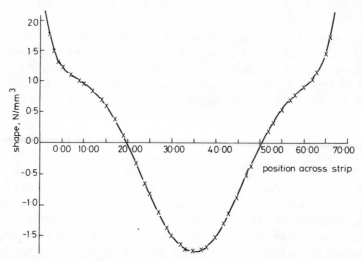

Fig. 4.8 *Shape change due to a 1-mm movement in rack 5 (for 1.7-m wide strip and rolls without camber)*

shape-operating point and these relate the shape changes to the As-U-Roll changes. If the shape is measured at eight points across the strip the gain matrix has the form (units = N/mm^3)

$$G_m = \begin{bmatrix} 4\cdot0 & 2\cdot0 & -0\cdot9 & -1\cdot7 & -1\cdot4 & -1\cdot1 & -1\cdot5 & -1\cdot0 \\ 1\cdot5 & 2\cdot2 & 1\cdot0 & -0\cdot8 & -0\cdot8 & -1\cdot1 & -1\cdot5 & -0\cdot9 \\ -0\cdot3 & 0\cdot9 & 2\cdot1 & 0\cdot4 & -0\cdot1 & -0\cdot9 & -1\cdot3 & -0\cdot7 \\ -0\cdot9 & -0\cdot5 & 1\cdot4 & 1\cdot6 & 1\cdot1 & 0\cdot1 & -1\cdot2 & -0\cdot7 \\ -0\cdot9 & -1\cdot2 & -0\cdot3 & 1\cdot5 & 1\cdot5 & 1\cdot1 & -0\cdot6 & -0\cdot8 \\ -0\cdot8 & -1\cdot3 & -1\cdot3 & 0\cdot2 & 0\cdot7 & 1\cdot4 & 0\cdot9 & -0\cdot4 \\ -0\cdot8 & -1\cdot3 & -1\cdot3 & -0\cdot8 & -0\cdot5 & 0\cdot7 & 2\cdot3 & 1\cdot2 \\ -0\cdot6 & -1\cdot4 & -1\cdot4 & -1\cdot4 & -1\cdot3 & -0\cdot6 & 2\cdot0 & 4\cdot0 \end{bmatrix}$$

$$(4.27)$$

The negative gains in this matrix result from the assumption that the average tension is maintained constant.

The gains include small errors due to numerical problems and due to the fact that the mill is non-linear. The mill gains are calculated using the usual small change procedure. The gains are dependent upon the operating point and are very dependent upon the strip width. Note that the above gain matrix is symmetric approximately and this is always the case if the strip is centred in the mill.

4.4 The dynamic mill model

4.4.1 Introduction
The mathematical model for the system will be described first in the following sections. This model is represented in state equation form which is convenient for both analysis and simulation. The dynamic model simulation is then discussed. This simulation is required for testing the proposed closed-loop shape control system and for tuning the controller under different rolling conditions.

4.4.2 The state space system description
The various sub-systems (Fig. 4.9) which form the shape control system are described below in state-equation form. The back-up roll actuators (As-U-Roll) are assumed non-interacting and are represented by second-order systems consisting of an integrator and time-constant (Appendix A) in cascade. There is also a dead zone arising from the solenoid valve spool and from friction in the hydraulic motor but this will be neglected since the effect can be mitigated by placing a position control loop around the shape actuators. In either case the shape actuators may be represented by the following

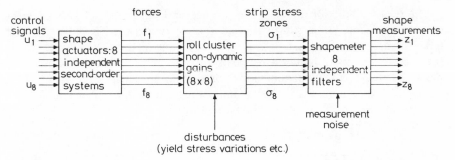

Fig. 4.9 *Block diagram of a Sendzimir mill roll-bending shape-control system*

state-space equations:

$$\dot{x}_a(t) = \Lambda_a x_a(t) + B_a u(t) \tag{4.28}$$

$$y_a(t) = C_a x_a(t) \tag{4.29}$$

where $x_a(t)$, $u(t) \in R^8$. The matrices (C_a, Λ_a, B_a) are diagonal and do not depend upon the material being rolled (schedule independent). There is some non-linear interaction between the actuators but this may be neglected since it can be eliminated, if necessary, by stiffening the hydraulic supplies.

The mill cluster is assumed to be non-dynamic so that the relationship between an actuator change $\delta x_a(t)$ and a shape-profile change $\delta x_m(t)$, at the roll gap, is simply a gain which may be calculated using the static model. The mill equation becomes:

$$y_m(t) = G_m y_a(t) + v_m(t) \tag{4.30}$$

where $y_m(t) \in R^{n_m}$. The number of outputs n_m is taken to be equal to the number of points across the strip width to be represented. For simplicity n_m is chosen as $n_m = 8$ in the dynamic model but this need not correspond with the value of n_m used in the control system design. The gain matrix G_m depends upon the material being rolled, although the matrix is approximately constant during a given pass of the reversing mill. Shape profiles are often parameterized to simplify control design studies and in such cases the matrix G_m is a function of a set of parameters p. These parameters may be calculated easily for a given shape profile. The matrix $G_m(p)$ may be calculated using the static model or using on-line measurements. The vector $v_m(t)$ represents shape disturbances which may result from changes in the input shape profile or indirectly via changes in the input gauge profile, material hardness or thermal camber.

Interaction between the automatic gauge control system (time constant $\simeq$ 30 ms) and the shape-control system may be neglected since the screwdown system on a Z-mill does not influence the strip shape. Recall that the screwdown eccentrics act on each back-up roll saddle, so that roll-bending does not occur as in four-high mills.

The transfer function between the roll gap and the shapemeter is a subject of current debate. It is suggested that this be represented as either a pure delay or a simple time lag.[21, 22] An investigation of the mechanisms which produce this transfer function reveals that both effects are present, to some degree. Thus, in the dynamic model simulation the strip transfer function (Appendix B) includes a time delay and a simple lag term in cascade.

The state equation for the strip has the form

$$\dot{\mathbf{x}}_s(t) = \Lambda_s \mathbf{x}_s(t) + B_s \mathbf{y}_m(t)$$

$$\mathbf{y}_s(t) = C_s \mathbf{x}_s(t) \tag{4.31}$$

$$\mathbf{y}_s'(t) = \mathbf{y}_s(t - \tau)$$

where Λ_s and B_s are square ($n_m \times n_m$), diagonal and depend upon the speed of the mill and the strip description. These matrices are therefore schedule dependent. It is sometimes convenient to model the strip by a second-order system rather than a lag and time delay in cascade. The corresponding transfer function may then represent either a simple lag or a Padé approximation to a transport delay or a combination of both elements. This transfer function has been identified from simple step-response tests on the mill.

The shapemeter (Section 4.2.3) forms the output subsystem and may be represented by a number of independent second-order transfer functions (Appendix C). The noise on the shape measurements is sinusoidal and proportional to the speed of the measuring roll and also contains a wide-band component. This noise signal is represented by the vector $\mathbf{v}_0(t)$ and is assumed to be a white-noise signal plus a coloured-noise signal with spectrum $\Phi_0/(s^2 + \omega_0^2)$. The state equations for the shapemeter have the form

$$\dot{\mathbf{x}}_0(t) = \Lambda_0 \mathbf{x}_0(t) + B_0 \mathbf{y}_s'(t) \tag{4.32}$$

$$\mathbf{z}_0(t) = C_0 \mathbf{x}_0(t) + \mathbf{v}_0(t) \tag{4.33}$$

The matrices Λ_0, B_0 and C_0 are dependent upon the speed of the mill and are switched by the shapemeter electronics.

The previous state equations may be combined into the form

$$\dot{\mathbf{x}}(t) = A_1 \mathbf{x}(t) + A_2 \mathbf{x}(t - \tau) + B\mathbf{u}(t) + D\mathbf{v}_m(t) \tag{4.34}$$

$$\mathbf{z}_0(t) = C\mathbf{x}(t) + \mathbf{v}_0(t) \tag{4.35}$$

where

$$A_1 = \begin{bmatrix} \Lambda_a & 0 & 0 \\ B_s G_m(\mathbf{p})C_a & \Lambda_s & 0 \\ 0 & 0 & \Lambda_0 \end{bmatrix} \qquad A_2 = \begin{bmatrix} 0 & 0 & 0 \\ 0 & 0 & 0 \\ 0 & B_0 C_s & 0 \end{bmatrix} \tag{4.36}$$

$$B = \begin{bmatrix} B_a \\ 0 \\ 0 \end{bmatrix} \qquad C = [0 \quad 0 \quad C_0] \qquad D = \begin{bmatrix} 0 \\ B_s \\ 0 \end{bmatrix} \tag{4.37}$$

The above equation is a differential-difference equation which has been investigated from many points of view in the literature. However, most results are much more complicated than for the equivalent delay-free system. Thus, since the strip transfer function is not a pure delay, it is expedient and perhaps more realistic to model the delay element by the transfer-function approximation discussed above.

By approximating the delay the equations to be used for control design become

$$\dot{\mathbf{x}}(t) = A\mathbf{x}(t) + B\mathbf{u}(t) + D\omega(t) \tag{4.38}$$

$$\mathbf{z}_0(t) = C\mathbf{x}(t) + \mathbf{v}_0(t) \tag{4.39}$$

where

$$A \triangleq \begin{bmatrix} \Lambda_a & 0 & 0 \\ B_s G_m(\mathbf{p})C_a & \Lambda_s & 0 \\ 0 & B_0 C_s & \Lambda_0 \end{bmatrix} \tag{4.40}$$

and $\mathbf{x}(t) = [\mathbf{x}_a^T(t), \mathbf{x}_s^T(t), \mathbf{x}_0^T(t)]$, $\omega(t) = \mathbf{v}_m(t)$. The A matrix is lower triangular which may allow some computational simplifications. The Markov parameters for the system are given by $\tilde{M}_0 = CB = 0$, $\tilde{M}_1 = CAB = 0$ and $\tilde{M}_2 = CA^2B = C_0 B_0 C_s B_s G_m(\mathbf{p})C_a B_a$. This matrix is full rank if the mill matrix $G_m(\mathbf{p})$ is full rank. For simplicity the vectors $\omega(t)$ and $\mathbf{v}_0(t)$ are assumed to be white-noise signals with covariance matrices which may be specified from test data. However, in practice both signals will contain narrow-band frequency components which can be allowed for in the control system design.

4.4.3 The dynamic model
The shape-control system involves a multi-input, multioutput plant and the shape-control system design[23] cannot be used with confidence without prior testing. The dynamic model enables various controller designs to be compared and verified. This model uses information obtained both from the static model and from plant tests. The model is also used to tune the shape controller to obtain the best transient responses. The dynamic model simulation is in the form of a FORTRAN computer program.

The time response of the system described in the previous section may be calculated using

$$\mathbf{x}(t) = \Phi(t - t_0)\mathbf{x}(t_0) + \int_{t_0}^{t} \Phi(t - \tau)B\mathbf{u}(\tau)\, d\tau \tag{4.41}$$

where $\Phi(t)$ is the state transition matrix $\Phi(t) \triangleq \exp(At)$. This equation is discretized[31] for use in the dynamic model computer simulation and has the form

$$\mathbf{x}(n + 1)T = \phi(T)\mathbf{x}(nT) + \Delta(T)\mathbf{u}(nT) \tag{4.42}$$

where $\phi(T) \triangleq \exp(AT)$ and $\Delta(T) \triangleq \int_0^T \Phi(T - s)B\, ds$. Equation (4.42) follows from (4.41) by assuming $\mathbf{u}(t)$ is constant during the integration intervals $t_k - t_{k-1}$.

The dynamic model is represented by four interconnected subsystems: (1) the crown adjustment actuators, (2) the mill cluster, (3) the strip dynamics and (4) the shapemeter subsystem. The main program is illustrated in Fig. 4.10 and this controls the inputs and outputs of the main subprograms

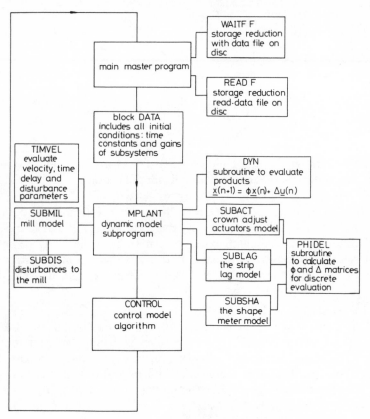

Fig. 4.10 *Dynamic model flow diagram*

MPLANT and **CONTROL** which represent the dynamic shape model and the controller model, respectively. The data needed for the simulation of the physical system, that is time constants, gains, schedule constants and disturbances, are stored in block data.

The calculation of the time response of the system is based upon the state transition equations above. The various steps executed by the program at each step length are illustrated in Fig. 4.11. The actuator subsystem inputs are either given in block data, for the open-loop case, or generated by the controller for the closed-loop case. For each time instant t_i the input to a subsystem is the output of the previous subsystem.

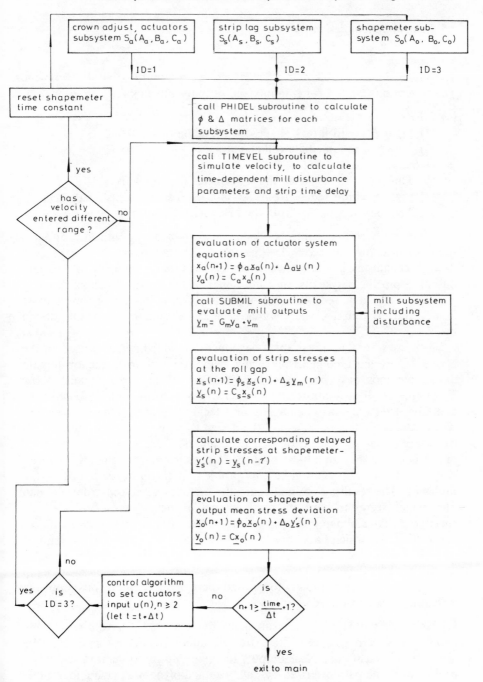

Fig. 4.11 *Dynamic model flow chart showing calculations for each step*

4.5 Control-system design

4.5.1 Introduction

The shape control problem on a Sendzimir mill is a multivariariable problem. The system has the following inputs and outputs affecting strip shape:

(a) *Inputs*
   (1) Eight As-U-Roll rack shape control actuator inputs.
   (2) Two first intermediate roll actuators.

(b) *Outputs*
   (1) Thirty-one signals from the shapemeter (not all active).
   (2) Eight position signals from the As-U-Roll shape-adjustment racks.
   (3) Two-position signals from the first intermediate rolls.

There are also certain controls which have a secondary influence on strip shape such as the side eccentrics (Section 4.2.2), the upper and lower screw-down mechanisms and the tension control and gauge controls. Inputs and outputs from these systems plus additional schedule data (such as the strip width) are also required. The controller will therefore have a minimum of 41 inputs and 10 outputs. Clearly this is a challenging multivariable design problem.

There are several multivariable design methods which might be used for the above design and a comparison is now being made. Two frequency-domain design methods have been considered to date; the Inverse Nyquist Array Technique due to Rosenbrock[24] and the two-stage (Multivariable Root Locus and Characteristic Locus) compensation technique due to MacFarlane and Kouvaritakis.[25] However, most of the design effort is aimed at a more classical approach at shape-control system design. The difficulty with the above methods is that they do not cater for the stochastic nature of the disturbances. An optimal control solution to this problem is therefore described in the following. The resulting optimal controller includes a reduced-order Kalman filter which seems to have many advantages in such systems.[26] It is a little surprising that Kalman filters have not been employed previously in shape-control systems which have very noisy operating environments. However, this may simply be due to the lack of previous work on the closed-loop shape-control problem.

Whichever design method is to be used there are several obvious problems which must be considered:

(a) How can the 31 shape measurements be combined to give a reasonable number of system outputs? There are two alternatives: the signal might be grouped into eight equivalent outputs to square up the system; or the shape profile might be parameterized by, say, a cubic polynomial, giving four effective outputs.

(b) How might the movement between the adjacent shape actuators (As-U-Rolls) be limited? The difference in the rack positions of adjacent actuators must be limited due to mechanical stress limits imposed by the mill manufacturers.

(c) How should the first intermediate rolls be controlled? A two-level control strategy seems desirable where the As-U-Rolls are used continuously and the first intermediate rolls are used intermittently. The second level of control (first intermediate roll control) will not be considered further here.

(d) Certain shape profiles must be avoided and thus constraints on the control system are necessary so that the system avoids these operating regions, even transiently.

4.5.2 The optimal control performance criterion

In many applications of optimal control theory there are difficulties in specifying the integral performance criterion which is constructed rather artificially. The shape-control problem does not share this difficulty and in fact there is a natural performance criterion which is obtained from the performance requirements. First consider the second (b) of the design problems mentioned in the previous section. The position difference between two adjacent actuators must be limited so that the integrand of the cost function must include a term

$$I_1 = \sum_{i=1}^{7} (u_i(t) - u_{i+1}(t))^2 m_{i,\,i+1} \qquad (m_{i,\,i+1} > 0) \tag{4.43}$$

A reference position for the actuators $(\mathbf{u}_0(t))$ is also to be specified, so that a second term to be minimized becomes

$$I_2 = \sum_{i=1}^{8} (u_i(t)) - u_{i0}(t))^2 m_i \qquad (m_i > 0) \tag{4.44}$$

To express these terms as a matrix quadratic form note that

$$(u_i - u_j)^2 = (u_i - u_{i0} - (u_j - u_{j0}))^2$$
$$= (u_i - u_{i0})^2 + (u_j - u_{j0})^2 - 2(u_i - u_{i0})(u_j - u_{j0})$$

thus

$$I_1 + I_2 = (\mathbf{u}(t) - \mathbf{u}_0(t))^T R(\mathbf{u}(t) - \mathbf{u}_0(t)) \tag{4.45}$$

and if $n_i \triangleq m_i + m_{i-1,\,i} + m_{i,\,i+1}$ then

$$R = \begin{bmatrix}
m_1 + m_{12} & -m_{12} & 0 & 0 & 0 & 0 & 0 & 0 \\
-m_{12} & n_2 & -m_{23} & 0 & 0 & 0 & 0 & 0 \\
0 & -m_{23} & n_3 & -m_{34} & 0 & 0 & 0 & 0 \\
0 & 0 & -m_{34} & n_4 & -m_{45} & 0 & 0 & 0 \\
0 & 0 & 0 & -m_{45} & n_5 & -m_{56} & 0 & 0 \\
0 & 0 & 0 & 0 & -m_{56} & n_6 & -m_{67} & 0 \\
0 & 0 & 0 & 0 & 0 & -m_{67} & n_7 & -m_{78} \\
0 & 0 & 0 & 0 & 0 & 0 & -m_{78} & m_8 + m_{78}
\end{bmatrix}$$

$$\tag{4.46}$$

This matrix has the simple triple-diagonal form and if for simplicity $m_i = m_{i,j} = 1$ for all $1 \le i, j \le 8$ then

$$
R = \begin{bmatrix}
2 & -1 & 0 & 0 & 0 & 0 & 0 & 0 \\
-1 & 3 & -1 & 0 & 0 & 0 & 0 & 0 \\
0 & -1 & 3 & -1 & 0 & 0 & 0 & 0 \\
0 & 0 & -1 & 3 & -1 & 0 & 0 & 0 \\
0 & 0 & 0 & -1 & 3 & -1 & 0 & 0 \\
0 & 0 & 0 & 0 & -1 & 3 & -1 & 0 \\
0 & 0 & 0 & 0 & 0 & -1 & 3 & -1 \\
0 & 0 & 0 & 0 & 0 & 0 & -1 & 2
\end{bmatrix}
\tag{4.47}
$$

The state-weighting matrix Q may be obtained by noting that the magnitude of each state-error term $e(t) = x(t) - x_r$ must be limited where x_r is a fixed reference level. This latter vector specifies the desired shape profile and the set points for the remaining state variables. The integrand of the performance criterion must therefore include a term of the form

$$
I_3 = (x(t) - x_r)^T Q (x(t) - x_r)
\tag{4.48}
$$

where Q is a diagonal matrix. As a rough guide the weighting elements $\{q_{is}\}$, associated with the shape-state variables, should be ten times larger than those associated with the remaining state variables. Since the strip edges are important increased weighting should be attached to the appropriate $\{q_{is}\}$ elements.

Because the system includes noise terms the performance criterion must include the ensemble averaging operator, or expectation operator E

$$
J(u) \triangleq E \left\{ \lim_{T \to \infty} \frac{1}{2T} \int_{-T}^{T} ((u - u_0)^T \mu^2 R (u - u_0) + (x - x_r)^T Q (x - x_r)) \, dt \right\}
\tag{4.49}
$$

The scalar μ allows the weighting between the control and state-error terms to be varied. This scalar is chosen by calculating the transient response of the system for various values of μ and by then selecting the response with the required dominant shape-time constant. An integral performance criterion is valuable in assessing the performance of the shape-control system. This assessment is particularly difficult in shape-control systems since profiles, rather than independent variables, are to be controlled.

4.5.3 The shape controller

The optimal shape controller may now be determined for the system defined in Eqs. (4.38) and (4.39) and the performance criterion defined above. Recall that the plant has 16 actuator states, 16 strip states and 8 shapemeter states (assuming 8 effective outputs). In the conventional solution to the stochastic optimal control problem all outputs are assumed contaminated by noise and the controller involves a Kalman filter of the same order as that of the plant

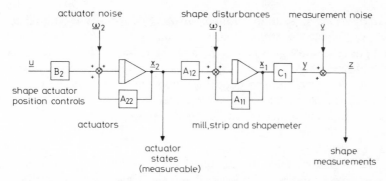

Fig. 4.12   *Open loop Sendzimir mill and shape-measurement subsystems*

(40 state variables). A controller of this size and complexity would be unrealistic. Fortunately, the problem may be redefined so that a smaller controller may be obtained.

Assume that the plant states in Eqs. (4.38) and (4.39) are grouped into two sets (by reordering) so that the system is separated into the input and output subsystems shown in Fig. 4.12. Note that all the states associated with the input, shape actuator and subsystem are measurable and that these measurements may be assumed noise free. None of the remaining state variables may be measured directly without the presence of noise. The solution[26] of this non-standard stochastic optimal control problem is illustrated in Fig. 4.13. Notice that the Kalman filter which is required has an order determined by the strip and the shapemeter (24 state variables) only. The optimal control feedback gain matrix

$$K_c = [K_{c1} \quad K_{c2}] \tag{4.50}$$

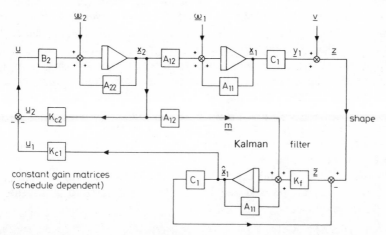

Fig. 4.13   *A combined state and state-estimate feedback solution to the shape-control problem*

is obtained from the solution of the equivalent deterministic optimal control problem by solving the algebraic Riccati equation.[27]

The optimal control gain matrix K_c and the Kalman filter gain matrix K_f, may need to be changed with line speed. The strip and shapemeter subsystems in the Kalman filter are also dependent upon the line speed and it may be necessary to change the elements of the matrices (C_1, A_{11}) as the shapemeter switches its time constants, in various speed ranges.

4.5.4 Shape-profile parameterization

It is usual to parameterize the measured shape profile so that the number of variables to be controlled is reduced. Thus, the shape profile might be represented by a third-order polynomial where X is the distance across the strip. Higher order bending can be neglected because of the mechanical construction of the mill. The coefficients of the polynomial may be found using linear regression theory.[28] The problem is to calculate the coefficients of the polynomial $aX^3 + bX^2 + cX + d$ given the $j_s(j_s \leq 31)$ measurements of strip shape Y_i at each point X_i. In matrix form

$$Y = XZ_1 + \epsilon \tag{4.51}$$

where ϵ is a vector of errors and $Z_1^T \triangleq [a \; b \; c \; d]$. The matrix X is known and thus the estimate $\hat{Z}_1$ which minimizes the error sum of squares $\epsilon^T \epsilon$ is

$$\hat{Z}_1 = (X^T X)^{-1} X^T Y \tag{4.52}$$

Note the above result does not depend upon the distribution of the errors but it is necessary to assume the errors are normally distributed if statistical tests or confidence limits are used. The matrix $M \triangleq (X^T X)^{-1} X^T$ is constant for a given strip width.

Using the above approach the j_s shape measurements from the shapemeter are reduced to four parameters $\{\hat{a}, \hat{b}, \hat{c}, \hat{d}\}$. This is equivalent to transforming the outputs from the plant by the constant matrix M. Clearly the Kalman filter must include this transformation. Now it is desirable to limit the size of the filter and thus the question arises as to whether it is necessary to model eight sets of strip and shapemeter states. To enable four parameters to be estimated at least four sets of outputs are required. Thus, to economize on the size of the filter the strip and shapemeter are represented by four sets of state variables (recall that the number of points across the strip which are treated as outputs is arbitrary). In this case the above transformation matrix M_1 is square and is non-singular.

The state variables and outputs associated with the shapemeter may be transformed, using $x_0(t) = M_0 x_1(t)$ and $z_1(t) = M_0^{-1} z_0(t)$, so that Eqns. (4.32 and (4.33) become

$$\dot{x}_1(t) = M_0^{-1} \Lambda_0 M_0 x_1(t) + M_0^{-1} B_0 y_s'(t) \tag{4.53}$$

$$z_1(t) = M_0^{-1} C_0 M_0 x_1(t) + v_1(t) \tag{4.54}$$

where $v_1(t) \triangleq M_0^{-1} v_0(t)$. The system matrices, for the augmented system (4.38) and (4.39), are as given in (4.36) and (4.37) but (A, C) become

$$A = \begin{bmatrix} \Lambda_a & 0 & 0 \\ B_s G_m(\mathbf{p}) C_a & \Lambda_s & 0 \\ 0 & M_0^{-1} B_0 C_s & M_0^{-1} \Lambda_0 M_0 \end{bmatrix} \tag{4.55}$$

$$C = [0 \quad 0 \quad M_0^{-1} C_0 M_0] \tag{4.56}$$

In the above representation the observations signal $z_1(t)$ which is input to the filter is the parameter vector $z_1(t) = M_0^{-1} z_0(t)$. The filter includes state variables representing both the strip shape $x_s(t)$ and the parameters $x_1(t)$, and the control feedback gain matrix K_c may be partitioned into shape and parameter feedback blocks. This scheme has the advantage that either the strip shape or the parameters (as in the classical approach) may be treated as primary control variables. The relative importance of the two is determined by the weighting attached to the appropriate state variables in the Q matrix.

The Kalman filter for the above system has twelve states only. This compares favourably with the order of a compensator designed using classical techniques.

The Kalman filtering control scheme has the following advantages:

(1) Disturbance inputs to the system may be represented by a colouring filter driven by white noise and these states may be estimated using the filter. Thus, valuable information on the incoming shape profile might be obtained.
(2) Additional filtering of the shape signals is provided automatically.
(3) The filter may be tested using an open-loop control to verify the only dynamic section of the controller.
(4) The shapemeter narrow-band noise may be modelled as in (1) to increase the effective filtering action.

4.6 Concluding remarks

The design of closed-loop shape-control systems is a relatively new problem and thus previous experience cannot be invoked when designing these systems. In this situation the development of models for the various mill subsystems assumes added importance. It is surprising that there has been no previous published work (excepting our own) on the development of a static model for a Sendzimir mill. This model is required to provide the mill gains but it also has the more important function of explaining the relatively complex behaviour of the system.[30]

Due to a lack of published work in this area there is even difficulty in defining the dynamic model of the mill. There is, for example, a current debate as to the best way to represent the strip transfer function. Identification methods are helpful in choosing the most appropriate representation of elements; however, since the rolling mills in question are used for rolling stainless steel, such tests can be expensive. Engineering a series of plant trials to maximize the gathering of data is also not a simple task. In such a complex system a dynamic model is clearly necessary to test the controller design under various mill operating conditions.[29]

The control design was not the main topic of the chapter and thus only a brief account was given of a possible design procedure. The Kalman filtering scheme could be made more sophisticated by using an extended Kalman filter to estimate the frequency of the sinewave disturbances. Alternatively, multivariable frequency domain design methods may provide a more conventional solution to the problem.

4.7 Acknowledgments

The shape-control problem is supported by the British Steel Corporation, GEC Electrical Projects Ltd. and the Science Research Council. I am grateful for the help and contributions from the following members of the project team: G. W. D. M. Gunawardene, M. Bourada, B. Coulbeck, G. F. Raggett and C. E. Carter (Sheffield City Polytechnic); K. Dutton and M. A. Foster (BSC); A. Thompson and A. C. Kidd (GEC).

4.8 Appendix A Actuator transfer function

The actuators may be represented by integrators together with some weak interaction between them, due to lack of stiffness in the hydraulics. For the present it is assumed that the actuators have position loops around them so that their transfer functions become

$$T_a(s) = \frac{8}{(1 + 0 \cdot 2 \, s)} \tag{4.57}$$

The state-space form of the actuator Eqns. (4.28) and (4.29) involves a block diagonal A matrix corresponding to the following equations for one actuator:

$$\begin{bmatrix} \dot{x}_1 \\ \dot{x}_2 \end{bmatrix} = \begin{bmatrix} 0 & 1 \\ -25 & -10 \end{bmatrix} \begin{bmatrix} x_1 \\ x_2 \end{bmatrix} + \begin{bmatrix} 0 \\ 200 \end{bmatrix} u \tag{4.58}$$

$$y = \begin{bmatrix} 1 & 0 \end{bmatrix} \begin{bmatrix} x_1 \\ x_2 \end{bmatrix} \tag{4.59}$$

4.9 Appendix B Strip transfer function

The time delay in the strip may be represented by either a first- or second-order Padé approximation for control design purposes. These have the following forms

First order

$$\frac{v(t - \tau)}{v(t)} = \frac{1 - p\tau/2}{1 + p\tau/2} \tag{4.60}$$

Second order

$$\frac{v(t - \tau)}{v(t)} = \frac{1 - p\tau/2 + p^2\tau^2/12}{1 + p\tau/2 + p^2\tau^2/12} \tag{4.61}$$

If a medium strip speed is considered, say 5 m/s, then since the distance from the shapemeter is 2·91 m the delay $\tau = 2\cdot91/5 = 0\cdot582$ s.

The strip time constant[21] (Grimble, 1976) may be calculated as follows:

$$\tau_1 = D/v = \frac{\text{distance between the coiler and roll gap}}{\text{strip speed}} \tag{4.62}$$

$$= 5\cdot32/5 = 1\cdot064 \text{ s} \tag{4.63}$$

The total strip transfer function may therefore be represented, using the first-order Padé approximation, in the form

$$T(s) = \frac{(1 - p\tau/2)}{(1 + p\tau/2)(1 + p\tau_1)} \tag{4.64}$$

where $\tau = 2\cdot91/v$ and $\tau_1 = 5\cdot32/v$. Note that each of the time constants is a function of the strip speed v. The state equations for the above subsystem, at $v = 5$ m/s, become

$$\begin{bmatrix} \dot{x}_1 \\ \dot{x}_2 \end{bmatrix} = \begin{bmatrix} 0 & 1 \\ -3\cdot2297 & -4\cdot3763 \end{bmatrix} \begin{bmatrix} x_1 \\ x_2 \end{bmatrix} + \begin{bmatrix} -0\cdot9398 \\ 7\cdot3427 \end{bmatrix} u \tag{4.65}$$

$$y = [1 \quad 0] \begin{bmatrix} x_1 \\ x_2 \end{bmatrix} \tag{4.66}$$

4.10 Appendix C The shapemeter transfer function

For a medium strip speed of less than 5 m/s the strip time-constant $= 0\cdot74$ s. The shapemeter may be modelled by the following second-order transfer function:

$$T_0(s) = \frac{1}{(1 + 0.74 \text{ s})(1 + 0.01 \text{ s})} \tag{4.67}$$

The shapemeter is approximated by the following transfer function:

$$T_0(s) = \frac{1}{(1 + 0.74 \text{ s})} \tag{4.68}$$

The fast mode is neglected in the present analysis. The shapemeter time constants are switched with line speed, so that the maximum ripple on the output signal is limited to 16%. The state equation for each shapemeter channel becomes $\dot{x} = -1.351x + 1.351u$ and $y = x$ for a line speed of 5 m/s.

4.11 References

1 EDWARDS, W. J.: 'Design of a cold-rolling mill thickness controller incorporating tension variation', *J. Inst. Met.*, March, 1975, pp. 59–67

2 PEARSON, W. K. J.: 'Shape measurement and control', *J. Inst. Met.*, 1964–65, **93**, pp. 169–178

3 SABATINI, B., WOODCOCK, J. W., and YEOMANS, K. A.: 'Shape regulation in flat rolling', *J. Iron Steel Inst.*, 1968, **206**, pp. 1214–1217

4 SABATINI, B., and YEOMANS, K. A.: 'An algebra of strip shape and its application to mill scheduling', *J. Iron Steel Inst.*, December, 1968, **206**, pp. 1207–1213

5 GUNAWARDENE, G. W. D. M., and GRIMBLE, M. J.: 'Development of a static model for a Sendzimir cold rolling mill', presented at IMACS (AICA) Symposium on Simulation of Control Systems, Technical University, Vienna, Austria, September, 1978

6 GUNAWARDENE, G. W. D. M., and GRIMBLE, M. J.: 'Development of a static model for a Sendzimir cold rolling mill', *Sheffield City Polytechnic Research Report*, No. EEE/13, April, 1978

7 EDWARDS, W. J., and SPOONER, P. D.: 'Analysis of strip shape', from 'Automation of tandem mills', G. F. Bryant (ed.), Iron and Steel Institute, 1973, p. 176

8 SIVILOTTI, O. G., DAVIES, W. E., HENZE, M., and DAHLE, O.: 'Asea-Alcan AFC System for cold rolling flat strip', *Iron & Steel Eng.*, June, 1973, pp. 83–90

9 KNOX, T. J.: 'Operating experience with the Loewy, Robertson Vidimon Shapemeter at Shotton Works', *Proc. Conf. Met. Soc.*, Chester, March, 1976, p. 55, Shape Control

10 DAHLE, O., SIVILOTTI, O. G., and DAVIES, W. E.: 'The "stressometer" shapemeter and automatic flatness control', Asea Vasteras, Sweden, pp. 123–128

11 HENZE, M.: 'Stressometer flatness transducer for cold strip mills', *ASEA J.*, **43**(6), pp. 127–129

12 URAYAMA, S., TAKATOKU, Y., NIWA, Y., and SAWADA, Y.: 'Experience in developing shape control for a Sendzimir mill', *Proc. Conf. Met. Soc.*, Chester, March, 1976, p. 101, Shape Control

13 GRIMBLE, M. J.: 'Shape control for rolled strip', *Inst. Mech. Eng.*, CME, 1975, pp. 91–93

14 DAVIES, N.: 'Measurement of strip, shape, speed and length', *Iron & Steel Eng.*, August, 1972, pp. 391–395

15 TIMOSHENKO, S., and GOODIER, J. N.: 'Theory of Electricity' (McGraw-Hill, 1961)

16 SPOONER, P. D., and BRYANT, G. F.: 'Analysis of shape and discussion of problems of scheduling set-up and shape control', *Proc. Conf. Met. Soc., Chester*, March 1976, p. 19, Shape control

17 GRIMBLE, M. J.: 'A roll force model for the cold rolling of thin sheet', Univ. of Birmingham Ph.D. Thesis, October, 1974

18 GRIMBLE, M. J., FULLER, M. A., and BRYANT, G. F.: 'A non-circular arc roll force model for cold rolling', *Int. J. Numer. Methods Eng.*, 1978, **12**, pp. 643–663

19 BRYANT, G. F., and OSBORN, R.: 'Derivation and assessment of simplified models for torque, slip and neutral angle', from 'Automation of tandem mills', G. F. Bryant (ed.), Iron and Steel Institute, 1973, p. 279

20 GOLTEN, J. W.: 'Analysis of Cold Rolling with Particular Reference to Roll Deformations', University of Wales, Swansea, Doctorial Dissertation, April, 1969

21 GRIMBLE, M. J.: 'Tension controls in strip processing lines', *Met. Technol.*, October, 1976, pp. 445–453

22 GRIMBLE, M. J.: 'Catenary controls in metal-strip processing lines', *GEC J. Sci. & Technol.*, 1974, **41**(1), pp. 21–26

23 PAWELSKI, O., SCHULER, V., and BERGER, B.: 'Development of a shape control system in cold strip rolling', *Proc. ICSTIS, Suppl. Trans., ISIJ*, 1971, **11**, pp. 692–704

24 ROSENBROCK, H. H.: 'Computer-aided control system design', (Academic Press, 1974)

25 MACFARLANE, A. G. J., and KOUVARITAKIS, B.: 'A design technique for linear multivariable feedback systems', *Int. J. Cont.* 1977, **25**(6), pp. 837–874

26 GRIMBLE, M. J.: 'A reduced order optimal controller for discrete-time stochastic systems', Research Report, Sheffield City Polytechnic, No. EEE/29/March 1979, to be published, *Proc. IEE*, 1980

27 KIRK, D. E.: 'Optimal control theory—An introduction' (Prentice-Hall, 1970)

28 DRAPER, N., and SMITH, H.: 'Applied regression analysis' (Wiley Inter-Science, 1966), p. 58

29 SHEPPARD, T., and ROBERTS, J. M.: 'Shape control and correction in strip and sheet', *Int. Metall. Rev.*, 1973, **18**, p. 171

30 O'CONNOR, H. W., and WEINSTEIN, A. S.: 'Shape and flatness in thin strip rolling', *J. Eng. Ind.*, November, 1972, pp. 1113–1123, *Trans. of the ASME*

31 KUO, B. C.: 'Discrete-data control systems' (Prentice-Hall, 1970)

Chapter 5

Representation and control of turbogenerators in electric power systems

B. W. Hogg

List of principal symbols

δ	rotor angle, $p = d/dt$
I_{fd}	field current
I_d, I_q	stator currents in d and q axis circuits respectively
V_b	busbar voltage
V_t	machine terminal voltage
P_t	power delivered at machine terminals
Q_t	reactive power delivered at machine terminals
T_e	airgap torque
T_M	generator shaft torque
X_T	transmission line reactance
R_T	transmission line resistance
R_m	stator resistance
I_{kd}, I_{kq}	damper circuit currents in d and q axes respectively
ψ	flux linkages
V_{fd}	field voltage
X_{ad}	stator-field mutual reactance
X_{fd}	field winding self-reactance
X_d, X_q	synchronous reactances in d and q axes
X_{kd}, X_{kq}	self-reactances of d and q axis damper circuits
ω_0	angular frequency of infinite busbar
U_M, U_I	actuating signal to governors on inlet and intercept valves respectively
G_V	position of inlet valves
τ_H, τ_I, τ_L	time constants associated with HP, IP and LP stages of turbine

τ_G, τ_C time constants of inlet and intercept valves
τ_{GL}, τ_T time constants of governors and turbine in linearized model
H inertia constant
T_R turbine reheat time constant
K_d mechanical damping coefficient
F_H, F_I, F_L fractions of total power from HP, IP and LP stages

5.1 Introduction

Modelling fulfils a vital role in investigations of the performance of electric power systems and development of new controllers for turbogenerators. Numerous mathematical models have been used[1-5] to describe these large and complex systems, and representations have become increasingly detailed, keeping pace with modern technology and new problems in power systems operation and control. Physical scaled models using small generators are also employed, with mathematical representation of the thermal plant.[5, 6] These serve for validation of theoretical models under all operating conditions, and provide a valuable means of gaining a clear understanding of practical aspects.

Most of the material in this chapter is the outcome of a project which is concerned with the development of multivariable controllers for large turbo-generator units. The availability of thyristor exciters, electrohydraulic governors and fast turbine valving has made it possible to implement rapid control, and it is therefore appropriate to seek improved performance by co-ordinating the functions of the exciter and governor in an integrated computer-control scheme. The development of suitable controllers involves extensive modelling and computer simulation of the boiler-turbine-generator system, both in the form of reduced-order linear models for control system design,[7, 8] and detailed non-linear models for evaluation of performance.[7, 8] In the initial stages of the project, only the generator and turbine were considered, and the outputs of computer simulations were compared with the results of laboratory tests on a micromachine system.[6-8] This is a physical simulation, with a scaled model generator, driven by a dc motor and controlled by a minicomputer.[6] The turbine was represented by analogue simulation in the motor control loop.

Recently, a more extensive turbine model has been adopted, and boiler dynamics have also been included. This additional complexity rendered the model unsuitable for analogue simulation, so it has now been transferred to the control minicomputer, which provides on-line real-time simulation as an integral part of the laboratory system.

Nearly all the controllers described in the literature have been designed for linearized analytical models of the plant. A turbogenerator may be represented by a non-linear set of equations, in which the coefficients depend on measured values of the system parameters. To design a controller, it is usual to reduce the order of the non-linear model, and linearize the equations by considering small deviations about a chosen steady-state operating condition. Models obtained in this manner may not provide an accurate representation of the plant.

An alternative approach is to apply the methods of system identification[9] to obtain models directly from the plant, either in transfer function,[10] state-variable,[11] or input-output[12, 13] format. This assumes that the plant is available, and is therefore of no value at the stage when the plant and its initial controller are being designed. However, if a computer-control scheme is used, installation and initial testing can proceed with controllers designed using analytical models which are known to provide satisfactory control. Subsequently, the plant can be identified, and the control software rearranged as necessary to achieve the best performance. This, in fact, is similar in principle to existing practices, in which controllers are tuned on-site when new plant is installed.

This chapter describes the simulation techniques used in the project, and discusses various aspects of the modelling. Results are presented which illustrate the performance of the system with state-space controllers, and compare the outputs of computer and physical simulations. It also describes the use of a recursive least-squares algorithm for on-line identification of a laboratory model turbogenerator. Pseudo-random-binary-sequence inputs are applied simultaneously to the exciter and turbine valving, and the responses of the turbogenerator outputs monitored by a computer. This provides the data required to obtain a state-variable model of the system. Comparisons of the responses of the laboratory system, identified models and various analytical models, show that identification produces much better low-order models, which are desirable for control-system design. Linear optimal controllers designed on the basis of identified models have been tested in the laboratory, and shown to perform well over a wide range of operating conditions.

5.2 Nonlinear mathematical model

5.2.1 Description of system
Initially this research was concerned with a single turbogenerator unit, connected to a large power system through a transformer and two parallel transmission lines, as shown in Fig. 5.1. The operating conditions of the unit are monitored and input to a computer, which determines the control signals for the steam valves and exciter. In practice, a distributed computer system would probably be used, based on mini- or microcomputers, with a hierarchical structure.[14] The model is not intended to represent a particular installa-

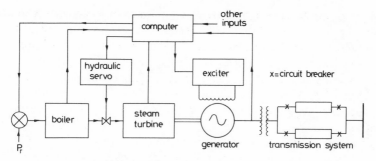

Fig. 5.1   *Schematic of turbogenerator system*

tion, but rather to be typical of large modern sets, so that general conclusions may be drawn from the results of the research programme.

There is, of course, a very wide range of plant and equipment available from a variety of manufacturers, in addition to which, the requirements and operating practices of power companies differ greatly, and it is impossible to devise a model which will satisfy everyone. It is necessary to take account of the purpose for which the model will be used, the extent and accuracy of available data, the precision required in the results, and the cost incurred by additional complexity in the computer simulation. The author has discussed these questions with engineers at manufacturers, power companies and universities, and the models described here are the present outcome of these deliberations. There will, no doubt, be further modifications as the research progresses.

5.2.2 Generator, exciter and transmission system

In this project, the requirement is for a generator model which is representative of large turbogenerators, but which can also be used to simulate a micro-alternator. This has a laminated rotor, with short-circuited damper windings, corresponding to the conventional mathematical model. The effect of magnetic saturation is neglected for a number of reasons, the most significant being the greatly increased complexity and consequent cost in computing time. However, it may be justified by the fact that the project is concerned with investigations of the performance of the same model with different control arrangements, and also that the laboratory model has a laminated rotor, thus reducing the effects of saturation. It could, however, be included by adjusting the mutual reactances with changes of operating conditions.[2]

The transmission system is represented as a lumped series resistance and reactance, with R_T being the combined resistance of the lines and transformer, and X_T including the transformer leakage reactance and line reactances.

The generator and transmission system equations appear below, and the symbols are defined in the List of principal symbols. The parameters used here are those of the microalternator system, which are listed in Appendix A.

$$p\psi_{fd} = \omega_0(e_{fd} - R_{fd}I_{fd}) \tag{5.1}$$

$$p\psi_d = \omega_0(V_b \sin \delta + I_d(R_m + R_T) + \psi_q) + \psi_q\dot{\delta} \tag{5.2}$$

$$p\psi_{kd} = \omega_0 R_{kd}I_{kd} \tag{5.3}$$

$$p\psi_q = \omega_0(V_b \cos \delta + I_q(R_m + R_T) - \psi_d) - \psi_d\dot{\delta} \tag{5.4}$$

$$p\dot{\delta} = \omega_0 \frac{T_M - T_e - K_d\dot{\delta}}{2H} \tag{5.5}$$

$$p\delta = \dot{\delta} \tag{5.6}$$

$$p\psi_{kq} = \omega_0 R_{kq}I_{kq} \tag{5.7}$$

$$\begin{bmatrix} I_{fd} \\ I_d \\ I_{kd} \\ I_q \\ I_{kq} \end{bmatrix} = \begin{bmatrix} X_{fd} & -X_{ad} & X_{ad} & 0 & 0 \\ X_{ad} & -(X_d + X_T) & X_{ad} & 0 & 0 \\ X_{ad} & -X_{ad} & X_{kd} & 0 & 0 \\ 0 & 0 & 0 & -(X_q + X_T) & X_{aq} \\ 0 & 0 & 0 & -X_{aq} & X_{kq} \end{bmatrix}^{-1} \begin{bmatrix} \psi_{fd} \\ \psi_d \\ \psi_{kd} \\ \psi_q \\ \psi_{kq} \end{bmatrix}$$

$$\tag{5.8}$$

$$T_e = \psi_d I_q - \psi_q I_d \tag{5.9}$$

$$V_{td} = V_b \sin \delta + R_T I_d - X_T I_q \tag{5.10}$$

$$V_{tq} = V_b \cos \delta + R_T I_q + X_T I_d \tag{5.11}$$

$$V_t = (V_{td}^2 + V_{tq}^2)^{1/2} \tag{5.12}$$

Various types of exciters have been used with large generators,[15] but only recently have thyristor exciters become available, which provide very fast control. When used in conventional mode, thyristor exciters require some form of stabilization,[16] but in the development of computer-control schemes this is assumed to be an integral part of the control strategy. Consequently, the exciter can be modelled very simply by a constant gain K_e and a single time constant τ_{ex}.

5.2.3 Boiler and turbine

The unit considered in these investigations is a two-pole turbogenerator, driven at 3000 rpm by a three-stage turbine with reheat. The steam is produced by a conventional coal- or oil-fired boiler and the steam flow controlled by both main and interceptor valves. The boiler and turbine constitute a very complex system, with many non-linearities,[17, 18] and it was initially decided to implement a relatively simple model, containing the essential features only.[7] The steam pressure at the turbine inlet valves was assumed to be constant, and the steam mass flow taken to be proportional to the valve position. Each stage was represented by a first-order transfer function, and the

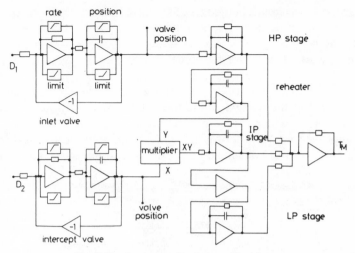

Fig. 5.2   *Analogue simulation of turbine*

outputs from each were weighted according to their relative contributions to the total shaft torque, and summed. The reheater was also described by a first-order transfer function, as were the valve servomechanisms. Limits were imposed on valve travel, and rates of movement. This model was made an integral part of the laboratory system by analogue simulation, and has been used in several experimental investigations.[7, 11]

The flow diagram is shown in Fig. 5.2, and the corresponding equations are

$$pY_H = \frac{G_M P_0 - Y_H}{\tau_H} \tag{5.13}$$

$$pY_R = \frac{Y_H - Y_R}{\tau_R} \tag{5.14}$$

$$pY_I = \frac{G_I Y_R - Y_I}{\tau_I} \tag{5.15}$$

$$pY_L = \frac{Y_I - Y_L}{\tau_L} \tag{5.16}$$

$$pG_M = \frac{U_M - G_M}{\tau_M} \tag{5.17}$$

$$pG_I = \frac{U_I - G_I}{\tau_I} \tag{5.18}$$

$$T_M = F_H Y_H + F_I Y_I + F_L Y_L \tag{5.19}$$

Y_H is the output of the high-pressure stage per unit, and $F_H Y_H$ is the contribution of the high-pressure stage to the total shaft torque T_M, etc.

Following the successful outcome of the initial research, it became apparent that a more elaborate description of the thermal plant was required to evaluate fully the control systems which had been developed, and provide a basis for further work. The improved model includes boiler dynamics, and takes account of non-linearities in the turbine and valving. These have received attention from many authors[18-20] and the aim here has been to devise a suitable model in the context of the present research, in which it constitutes only a part of the whole system model.

The improved model is shown in Fig. 5.3. The boiler is assumed to have a p.i.d. pressure control system, and both the fuel supply system and boiler are

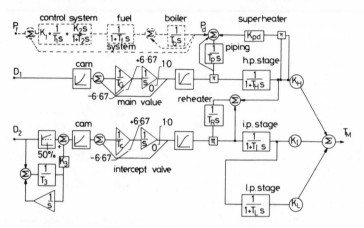

Fig. 5.3   *Boiler-turbine model*

represented by single time constants. The superheater is modelled as a simple flow resistance, having a pressure drop proportional to the square of the mass steam flow. The dynamic effect of the entrained steam in the piping connecting the boiler to the turbine is also accounted for by a time constant, and the simulation of the reheater has been modified. This model neglects losses, and is based on recognition that the main effect of the boiler on the turbine is due to the variation of pressure upstream of the main steam valve. The steam mass flow is a non-linear function of the valve positions, and this is usually corrected under steady-state conditions by a cam, or electronic function generator in the case of an electro-hydraulic governing system.[19] The interceptor valves may be continuously controlled, in which case constraints are necessary on valve movement.[19] However it is more usual to keep the interceptor valve fully open, except during emergency conditions. These effects have all been included in the model.

Parameter values have been obtained from published data and information provided by manufacturers. These are given in Appendix A, and appertain to an oil-fired system with electrohydraulic governors and fast turbine valving. The rate limits correspond to the fastest valve operation currently available.

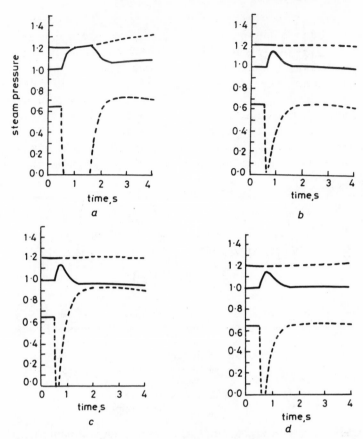

Fig. 5.4 *Steam pressure variations*
-------- down pressure
———— upstream pressure
-------- downstream pressure

The effect of valve movements on steam pressure may be observed by instantaneously reducing U_M to zero, and returning it to the initial steady-state value after time t. The consequent variations of steam pressures are shown in Fig. 5.4. In Fig. 5.4a, t is fairly long, and the boiler drum pressure increases after about 1·5 s. However, with small values of t, the drum pressure is virtually constant (Fig. 5.4b). Similar results are obtained when the demand signal is set to a new level after time t, as shown in Fig. 5.4c. Consequently, for short transient disturbances, the drum pressure can be taken as constant, and the part of the model shown by dotted lines can be neglected. Figure 5.4d was obtained using this reduced model, and is directly comparable with Fig. 5.4b. The two sets of curves are almost identical.

Simulation of a load rejection test is shown in Fig. 5.5, and the results are similar to those published elsewhere, making allowance for differences of

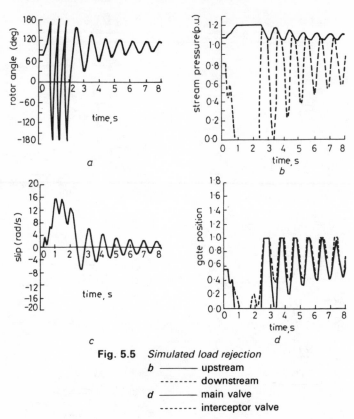

Fig. 5.5   *Simulated load rejection*

b ———— upstream
     -------- downstream
d ———— main valve
     -------- interceptor valve

system configuration. These results were obtained off-line, with the generator model defined in Eqns. (5.1) to (5.12), and a conventional speed governor with 2% droop.

5.3 Linearized mathematical models

The design of control systems is based on linearized models of the plant, usually in state-space or transfer-function format, and these models are normally of reduced order. In this case, the generator equations were reduced to either fifth or third order, and the turbine/governor to second order, assuming constant pressure at the main steam valve. This model was considered to be adequate within the limits imposed by linearization. Linearization is achieved by considering small perturbations about a selected operating point, defined in terms of the generated active and reactive power, and using a linear transformation to obtain a suitable set of measurable state variables.[21]

In deriving these equations from those given in Section 5.2.2, the effects of dampers have been ignored in the synchronous-machine equations, and it is assumed that the inlet and intercept valves are stroked simultaneously. The symbol Δ denotes the deviation from the steady state

$$\Delta p \delta = \Delta \dot{\delta} \tag{5.20}$$

$$p \, \Delta \psi_{fd} = \omega_0 \, \Delta V_{fd} - \omega_0 R_{fd} \, \Delta I_{fd} \tag{5.21}$$

$$p \, \Delta \psi_d = \omega_0 \, \Delta \psi_q + \omega_0 R \, \Delta I_d + \psi_q \, \Delta \dot{\delta} + \delta \, \Delta \psi_q + \omega_0 V_b \sin \Delta \delta \tag{5.22}$$

$$p \, \Delta \psi_q = -\omega_0 \, \Delta \psi_d + \omega_0 R \, \Delta I_q - \psi_d \, \Delta \dot{\delta}$$
$$- \delta \, \Delta \psi_d + \omega_0 V_b \cos \Delta \delta \tag{5.23}$$

$$p \, \Delta \delta = \frac{\Delta T_M \omega_0}{2H} - \frac{\Delta T_e \omega_0}{2H} - \frac{K_d \, \Delta \dot{\delta} \omega_0}{2H} \tag{5.24}$$

$$\Delta T_e = \psi_d \, \Delta I_q + I_q \, \Delta \psi_d - \psi_q \, \Delta I_d - I_d \, \Delta \psi_q \tag{5.25}$$

$$\Delta V_{bd} = V_b \cos \delta \, \Delta \delta \tag{5.26}$$

$$\Delta V_{bq} = -V_b \sin \delta \, \Delta \delta \tag{5.27}$$

$$\Delta \psi_{fd} = X_{fd} \, \Delta I_{fd} - X_{ad} \, \Delta I_d \tag{5.28}$$

$$p \, \Delta \psi_{fd} = X_{fd} p \, \Delta I_{fd} - X_{ad} p \, \Delta I_d \tag{5.29}$$

$$\Delta \psi_d = X_{ad} \, \Delta I_{fd} - (X_d + X_T) \, \Delta I_d \tag{5.30}$$

$$p \, \Delta \psi_d = X_{ad} p \, \Delta I_{fd} - (X_d + X_T) p \, \Delta I_d \tag{5.31}$$

$$\Delta \psi_q = -(X_q + X_T) \, \Delta I_q \tag{5.32}$$

$$p \, \Delta \psi_q = -(X_q + X_T) p \, \Delta I_q \tag{5.33}$$

$$p \, \Delta T_M = \frac{\{F_H + (F_I + F_L) U_{g0}\} \Delta G_V - T_M}{\tau_T} \tag{5.34}$$

$$p \, \Delta G_V = \frac{\Delta U_g - \Delta G_V}{\tau_G} \tag{5.35}$$

The linearized machine and turbine equations may be rewritten in matrix form as a seventh-order linear model, with state variables δ, $p\delta$, I_{fd}, I_d, I_q, G_V, T_M. Linear transformations of this model may now be obtained,[21] in order to replace I_d and I_q by measurable quantities. For example, a suitable state-variable set is

$$X^T = (\delta, \, p\delta, \, I_{fd}, \, P_t, \, V_t, \, G_V, \, T_M) \tag{5.36}$$

The model may be reduced to fifth order by assuming constant flux linkages in the generator, and increasing the rotor damping coefficient to compensate for the omission of damper windings.

5.4 Computer simulation

5.4.1 Requirements
The behaviour of the boiler/turbogenerator system with various control systems was investigated by computer simulation, over a wide range of operating conditions and circumstances. These include assessment of dynamic stability, which is done either by solving the non-linear equations or examining the eigenvalues of the closed-loop system using the linearized model. Transient performance following large disturbances must also be considered. These disturbances may be caused by short circuits on the transmission system, and the behaviour of the system is highly non-linear. Semi-permanent changes of system configuration may also arise, due to transformer tap changing or switching of transmission lines.

5.4.2 Computer program
The program is written in FORTRAN IV. It accepts a given operating point and generates the solution to the non-linear equations. A sequence of disturbances can be initiated in the form of a symmetrical three-phase short circuit at the sending end of the transmission line, followed by a step change of transmission line impedance to simulate the removal of a line to clear the fault. The fault duration and the prefault and postfault impedances are supplied to the program as data. The inputs may be held constant to represent an unregulated machine, or may be varied by a control system. The output is printed, with selected variables also displayed in graphical form, using a graph plotter attached to the computer.

Numerical methods are used to solve the equations. As is typical of power system models, the eigenvalues of the Jacobian, $\partial f/\partial x$, are widely separated, and the equations are classified as 'stiff'.[22] A numerical integration routine based on Gear's method,[23] which is able to cope with stiff systems, was selected from a numerical algorithm library. The integrating routine calls a user-supplied subroutine, inside which the derivatives of the state variables are calculated. The integration step size is automatically adjusted by the routine to meet a specified accuracy criterion.

The excitation voltage ceiling is imposed by testing the input V_{fd} and replacing it by its limiting value should the limit be exceeded. As the governor valve position signal may be a state variable rather than an input variable, the valve rate and position limits are simulated by modifying the calculated derivatives.

A symmetrical three-phase short-circuit at the sending end of the transmission line is effected by forcing the busbar voltage to zero, and reducing the values of R_T and X_T for the duration of the fault. At the end of the fault the busbar voltage is returned to its normal value and the post-fault transmission line impedance is inserted.

5.5 Multivariable controllers

5.5.1 Design considerations

The conventional analogue control of a turbogenerator is performed by an automatic voltage regulator (a.v.r.) and turbine governor. The basic function of these controllers is to maintain the terminal voltage of the generator and speed of the set at almost constant values during steady operating conditions, corresponding to specified reference levels. Usually, it is required to keep variations of terminal voltage within $\pm 0.5\%$, with a speed change of 4% from no load to full load. During transient conditions caused, for example, by a short circuit on the transmission system, the first requirement is to maintain stability by keeping the generator in synchronism with the power system. Once this has been achieved, the generator should return to normal running conditions as soon as possible. Under these circumstances, the behaviour of the system is highly non-linear.

Proposed multivariable controllers for turbogenerators may be separated into a number of categories as follows:

(*a*) Those which consider excitation control only.[24, 25]
(*b*) Those which seek to co-ordinate the control of exciter and turbine.[7, 26]
(*c*) Those which are concerned with transient conditions only.[26]
(*d*) Those which retain conventional governor and/or a.v.r. systems, and apply additional inputs to the setpoints.[8, 24–26]
(*e*) Those which replace the a.v.r. and/or governor by a multivariable controller.[7, 27] In this case, the basic steady-state requirements for controlling speed and/or terminal voltage must be satisfied by the multivariable controller.

Most of the controllers described in this chapter are in category (*e*). Various states are measured and are fed into a computer that contains the control algorithms, and supplies appropriate control signals to the exciter (V_{FD}) and turbine valving (U_g).

The usual procedure for designing a state-variable controller is to define a suitable linearized model of the system in the form of the well-known state equation for a system with n states and m inputs

$$\dot{x} = Ax + Bu \tag{5.37}$$

in which lower-case letters denote small deviations about steady-state conditions X, U.

Various techniques, such as the calculus of variations or dynamic programming, may be used to design a controller by optimizing a selected performance index.

By standard methods, a controller of the form

$$u = Kx \tag{5.38}$$

can be designed, where u and x are the deviations of the control and state vectors, respectively, from the reference values U and X, and K is the control matrix.

At this stage, a number of practical considerations arise which must be taken into account if the controller is to operate satisfactorily. These are:

(*a*) Although the state-variable set may be selected to contain only measurable quantities, it may not be easy to measure accurately the deviations of these quantities from specified reference levels.

(*b*) The generator is required to function over a wide range of steady operating conditions according to load demand, and it is essential that control be maintained while shifting from one to another. However, in these circumstances, the new reference vectors U and X are unknown, although they might be calculated on-line if the desired new operating conditions are known.

(*c*) Permanent changes of system configuration occur from time to time, for example, due to switching of transmission lines or on-load tap changing of transformers. This can cause undesirable offsets in some of the system variables, and the controller must be designed to ensure that offsets do not occur in those variables which would degrade system performance. Again, the new reference vectors are unknown, and in this case cannot be calculated.

(*d*) The controller must function during large transient disturbances, when the system behaviour is highly non-linear, and when there are constraints on the control inputs.

Some of the problems can be avoided or alleviated by retaining the normal a.v.r. and governor for steady-state control and considering transient conditions only,[26] or using as state variables only quantities which are zero in the steady state,[28] or backing off the steady-state values.[25] However, none of these methods seems to provide a satisfactory overall solution to the problems stated in (*a*) to (*d*) above. Another approach is described in Ref. 24, which involves a feedback-feedforward structure for excitation control using washout networks, but here also the performance is degraded when operating conditions change. All these methods employ standard state-space techniques for continuous systems, and are based on the measurement of deviations of state variables about reference values. Various means are then sought to achieve practical implementation of the controllers.

A new approach to the design of physically realizable controllers for turbogenerators takes account of the practical difficulties from the outset.[7, 27] These are specifically designed as discrete controllers using actual measured values of variables in place of deviations. Also, it is recognized that only some states have to be precisely controlled, and these are held at exactly the desired values by output integral control,[29] regardless of changes of operating conditions, at the expense of offsets on other variables which are of little significance. This structure enables a state-space controller to drive from one operating point to another, as required, and to cope with changes of system configuration.

5.5.2 Control algorithm

Integral action may be used to eliminate steady-state errors in as many output variables as there are control inputs, and the state equation (5.37) is therefore augmented to include integral terms, following the method described in Ref. 30.

Let Y be a vector containing the measured values of r selected outputs, at time t, and Y_D be a vector representing the corresponding demand levels, then

$$Y = CX = C(X_D + x) \tag{5.39}$$

where C is an $r \times n$ matrix consisting of elements from a unit matrix; therefore

$$Y - Y_D = Cx \tag{5.40}$$

Let

$$\dot{y} \approx Cx \tag{5.41}$$

Now Eqn. 5.37 can be augmented by Eqn. 5.41, to give the following model

$$\begin{bmatrix} \dot{x} \\ \dot{y} \end{bmatrix} = \begin{bmatrix} A & 0 \\ C & 0 \end{bmatrix} \begin{bmatrix} x \\ y \end{bmatrix} + \begin{bmatrix} B \\ 0 \end{bmatrix} u \tag{5.42}$$

or, in more compact notation,

$$\dot{Z} = \hat{A}Z + \hat{B}u \tag{5.43}$$

In discrete form this becomes

$$Z_{k+1} = \phi Z_k + \Delta U_k \tag{5.44}$$

where

$$Z_k^t = (x_k, y_k)$$

A controller may be designed by minimizing the quadratic performance index J using dynamic programming

$$J = \sum_{k=1}^{N} (x_{k+1}^t P x_{k+1} + y_{k+1}^t Q y_{k+1} + u_k^t R u_k) \tag{5.45}$$

For large values of N, the control law becomes

$$u_k = K Z_k \tag{5.46}$$

and partitioning the control matrix into submatrices gives

$$u_k = [K_p \mid K_I] \begin{bmatrix} x_k \\ y_k \end{bmatrix} \tag{5.47}$$

Substituting the actual values of variables instead of small deviations from the reference values[7]

$$u_k = U_k - U_{k-1}$$
$$x_k = X_k - X_{k-1} \tag{5.48}$$
$$y_k = Y_k - Y_D$$

The expression for y_k is valid when the sampling interval is small.
Then

$$U_k = K_p X_k + K_I \sum_{k=1}^{k} (Y_k - Y_D) \tag{5.49}$$

This is a discrete PI control algorithm based on measured values of variables (i.e. not deviations) at the latest sampling instant (k), whereas the controller described by Pullman and Hogg[7] uses information obtained at the previous sampling instant $(k - 1)$.

5.5.3 Controller design
Controllers have been designed for both fifth- and seventh-order linearized models,[7, 27] and, as an example, a fifth-order model is used here. The state-variable set is

$$x^t = (\delta, \, p\delta, \, V_t, \, G_V, \, T_M) \tag{5.50}$$

The first four states are directly measurable and the mechanical torque can be measured either explicitly,[16] or implicitly.[31]

With two inputs, only two output states can be selected for integral control. For practical reasons it is essential that terminal voltage be one of these. A possible second choice is power output, so that disturbances in the thermal plant are minimized and load dispatch on the power system is facilitated. Alternatively, if stability is likely to be a problem, it may be desirable to maintain the rotor angle at a constant value when conditions alter in the system.

Two sets of controllers have been designed, one with integral control on V_t and P_t, the other with integral control on V_t and δ. A dynamic programming algorithm was used to minimize the performance index (Eqn. 5.45), and a trial-and-error approach adopted to the selection of weighting matrices. Two controllers are considered here, which were designed at the nominal operating point $P_t = 0.8$ p.u., $Q_t = 0.2$ p.u. These are typical of the two types of controllers.

Controller 1

$$P_{\text{diag}} = 150, \, 120, \, 50, \, 5000, \, 200$$
$$Q_{\text{diag}} = 10, \, 15$$
$$R_{\text{diag}} = 1, \, 5$$

$$K_p = \begin{bmatrix} \delta & p\delta & V_t & G_V & T_i \\ -0{\cdot}205 & -0{\cdot}402 & 4{\cdot}033 & -4{\cdot}58 & -0{\cdot}146 \\ -0{\cdot}313 & 4{\cdot}38 & -139{\cdot}25 & 2{\cdot}243 & -1{\cdot}007 \end{bmatrix} \begin{matrix} u_g \\ v_{FD} \end{matrix}$$

$$K_I = \begin{bmatrix} \delta & V_t \\ -0{\cdot}092 & 0{\cdot}0059 \\ 1{\cdot}509 & -0{\cdot}364 \end{bmatrix}$$

Controller 2

$P_{\text{diag}} = 60, 30, 50, 4000, 300$

$Q_{\text{diag}} = 60, 30$

$R_{\text{diag}} = 1, 10$

$$K_p = \begin{bmatrix} p\delta & P_t & V_t & G_V & T_i \\ -0{\cdot}419 & -0{\cdot}2 & 3{\cdot}725 & 4{\cdot}437 & -0{\cdot}244 \\ 5{\cdot}040 & 2{\cdot}531 & -120{\cdot}814 & 2{\cdot}633 & -1{\cdot}865 \end{bmatrix} \begin{matrix} u_g \\ v_{FD} \end{matrix}$$

$$K_I = \begin{bmatrix} P_t & V_t \\ -0{\cdot}051 & -0{\cdot}0281 \\ 0{\cdot}802 & -0{\cdot}473 \end{bmatrix}$$

The performance of the turbogenerator with the above controllers was assessed by computer simulation, using the complete non-linear boiler-turbogenerator model, over a wide range of operating conditions, and found to be satisfactory.

5.6 Laboratory model

5.6.1 General description

A laboratory model turbogenerator was used to test the controllers developed by computer simulation. This establishes that the control algorithms can be implemented in real time, in the presence of noise, transducer limitations and other practical constraints, and give satisfactory performance. The laboratory system is shown in Fig. 5.6. It consists of a 3-kVA, 3-phase microalternator, which is a scaled model of a large synchronous generator, driven by a separately excited dc motor and connected to the laboratory busbar through a transformer and two parallel transmission lines.[6, 21] The turbine representations have already been described, and the output from the turbine simulation (T_M) controls the armature current of the dc motor. In Fig. 5.6, the boiler-turbine simulation is shown separately from the computer, corresponding to the original analogue arrangement, which is still available. With the new model, the thermal plant is simulated in the computer, which supplies T_M directly, via the hardware interface. The motor field current is held constant, so the shaft torque which drives the microalternator is proportional to $\dot{T}_M$.

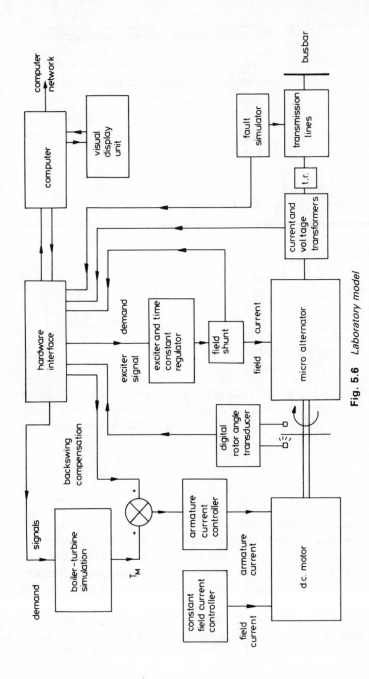

Fig. 5.6 *Laboratory model*

The microalternator is excited by a transistor power amplifier with negligible time constant, which also regulates the time constant of the field winding. The instrumentation,[32] provides inputs to a 20K minicomputer adjacent to the micromachine, via a CAMAC hardware interface. The computer contains the control algorithms, and supplies output signals for the exciter and turbine control. It is connected to a hierarchical computer network, with access to a mainframe computer where digital simulation studies are performed. This background processing capability is employed for program development, processing of test results, etc. A visual display terminal is used for system operation and to display test results. The transmission lines are simulated by RLC π-sections, with facilities for fault application, line switching etc.

When a loaded synchronous generator is subjected to a three-phase short circuit close to the terminals, the rotor initially moves against the direction of rotation before being accelerated in a forward direction. In a microalternator, this backswing can be very much greater than in a large turbogenerator, and therefore under these conditions it does not adequately simulate a full-scale generator.[33] The laboratory model includes compensation which ensures that the backswing is similar to that of a large machine.[34]

5.6.2 Turbine simulation

The original turbine model is simulated on an analogue computer, but the greater complexity of the subsequent model (Fig. 5.3), and particularly the large number of non-linearities, would cause difficulties in analogue simulation, so this model is simulated in the minicomputer, which provides on-line real-time simulation as part of the physical model.

The control and instrumentation system operates with a sampling interval of 20 ms, and the turbine simulation is updated with each sample. The computer program, which is written in Coral 66, accepts the demand signals to both main and intercept valves as input, and generates the solution of the model non-linear equations. Fast execution and accurate solutions are essential, and particular care was taken in selecting a suitable numerical technique, based on the Runge-Kutta-Merson algorithm. The non-integrable variables have been included within the solution routine to minimize errors. The program occupies 1K of core, and the execution time varies with operating conditions, up to a maximum of about 5 ms.

5.7 Implementation and test results

Controllers 1 and 2 were implemented on the laboratory model turbogenerator, and extensively tested. Both controllers function in a fully automatic mode, and can drive to a new operating point by means of a secondary controller,[7] or adjust to changes caused by switching operations on the transmission system. Satisfactory performance was obtained over the whole operating range of the generator, although the controllers are suboptimal for all but the nominal operating point at which they were designed. It was possible to design wide-range controllers which are also effective during transient conditions, and do not require any superimposed contingency action.[7]

In setting up the controllers, the generator was run independently, supplying a local load, and not synchronized to the laboratory busbars. The turbine droop characteristic is linearized by a simple algorithm which adjusts the power reference automatically, and the setting is specified or altered by typing the desired value into the computer. The linearized characteristic is shown in Fig. 5.7. The response to a step change of local load with controller 2 is

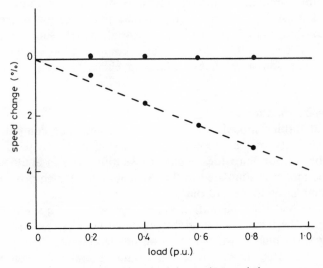

Fig. 5.7 *Linearized droop characteristic*

shown in Fig. 5.8. The instrumentation[32] was designed for use with the machine synchronized to the busbar, so these graphs show a substantial element of noise. However, the response of the system is considered to be satisfactory. The generator was then synchronized via the transmission system, retaining the local load on the machine, the aim being to maintain constant tie-line power as the local load changes. The system was subjected to step changes of local load, and the response with controller 2 is shown in Fig. 5.9.

The local load was removed, and dynamic stability boundaries obtained with both controllers. These are shown in Fig. 5.10.

The effects of a three-phase short circuit to earth at the sending end of one transmission line, which is cleared by switching out that line after 120 ms, are

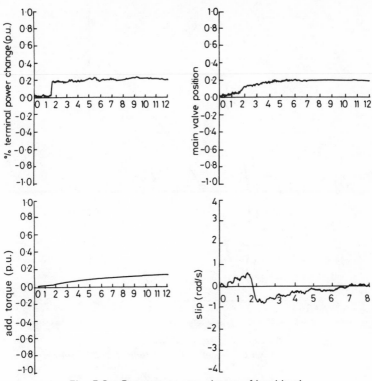

Fig. 5.8 *Response to step change of local load*

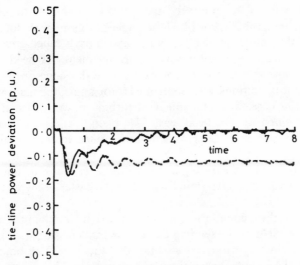

Fig. 5.9 *Control of tie-line power*
——— with controller
------- without controller

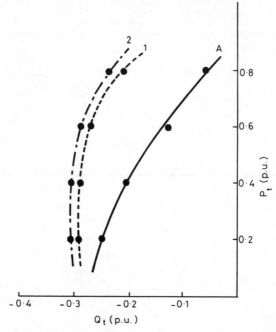

Fig. 5.10 Dynamic stability boundaries
A. No Controller; 1. Controller 1; 2. Controller 2.

shown in Figs. 5.11 and 5.12. In both cases, the terminal voltage recovers quickly, and returns to precisely the prefault level. With controller 1 (Fig. 5.11) the integral control on rotor angle brings it back to the prefault steady-state value, although the postfault impedance is much greater than the prefault impedance. A parallel governing arrangement has been assumed, so that initially both main and intercept valves have the same opening. To cater for the postfault conditions, and control of rotor angle, the main valve makes the most of the necessary adjustment of turbine power, since the intercept valve is constrained to remain at least 50% open during steady operation.[19] This does not occur when integral action is applied to the power output by controller 2 (Fig. 5.12). The controllers have been designed so that the valves move at the maximum rate during large disturbances, but only one close-open cycle is permitted. The rate limits used are the fastest now available from manufacturers.

Transient stability boundaries were obtained by applying a three-phase short circuit to earth at the sending end of one transmission line, and clearing the fault after time t_c without switching out the line. The boundaries are shown in Fig. 5.13, and illustrate the effectiveness of the controllers in this respect.

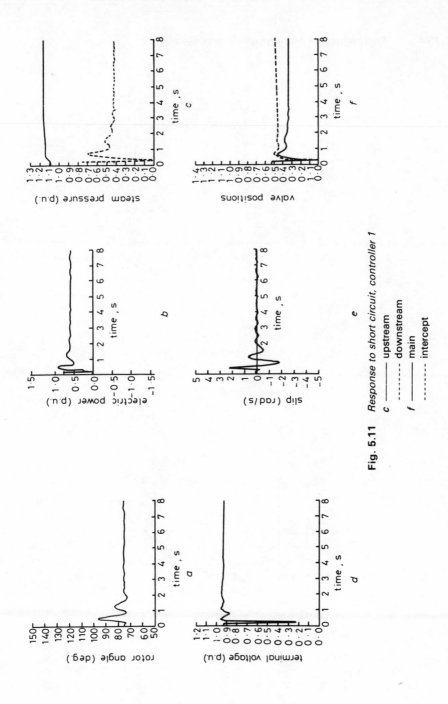

Fig. 5.11 Response to short circuit, controller 1

c ——— upstream
———— downstream
f ——— main
·········· intercept

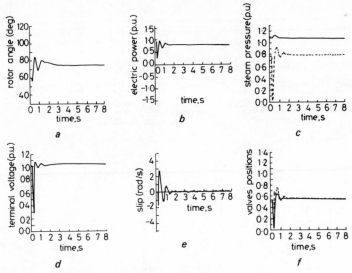

Fig. 5.12   *Response to short circuit, controller 2*

c ———— upstream
‑‑‑‑‑‑‑‑ downstream
f ———— main
‑‑‑‑‑‑‑‑ intercept

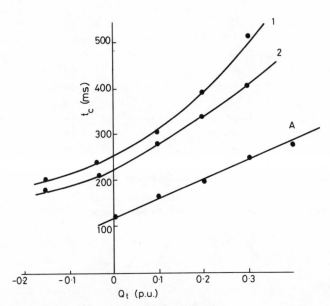

Fig. 5.13   *Transient stability boundaries*
A. No controller;   1. Controller 1;   2. Controller 2.

5.8 Identified linear models

5.8.1 Introduction

An alternative method of obtaining linear models is to apply the techniques of system identification.[9] The structure or order of the model must be selected with care, and the identification procedure then provides the values of system parameters. A recursive least-squares algorithm has been used for on-line identification of the laboratory model turbogenerator, in order to provide comparisons with linearized analytical models, and particularly to seek accurate low-order models which facilitate control system design. In the context of a computer-control scheme, this approach has the advantage that the model which is obtained by the computer includes the effects due to noise, transducer inaccuracies etc., and this model in turn is used to formulate the control algorithms, which then reside in the computer.

Three sets of linearized models have been used:

1. A thirteenth-order model based on the full set of Eqns. (5.1) to (5.19).
2. A ninth-order model, containing all the generator quantities, but with a second-order turbine representation.
3. A fifth-order model, in which the generator equations are reduced to third-order by assuming constant flux linkages and omitting damper windings. The state variables are

$$X^t = (\Delta V_t, \Delta I_t, \Delta p\delta, \Delta G_V, \Delta T_M)$$

5.8.2 Identification algorithm

Comparative studies of various recursive algorithms[35] have shown that the method of recursive least squares is one of the most efficient and reliable techniques for on-line identification, and is particularly economical in computing time and storage requirements. It also has good convergence characteristics, and can be applied to multivariable processes. The main disadvantage is that it may lead to biased estimates of identified parameters if the input data is corrupted by non-white noise, but this can be minimized by using a sufficiently large number of samples. Data is generated by applying small random disturbances to the plant inputs, and monitoring the responses of the states, one sample of information being input to the computer at each sampling instant.

The dynamics of a linear multivariable process may be represented in discrete form by the equation

$$X_k = \phi X_{k-1} + \Delta U_{k-1} \qquad (5.51)$$

in which the dimensions are

$$X = n \times 1 \qquad U = p \times 1 \qquad \phi = n \times n \qquad \Delta = n \times p$$

A vector Z_{k-1} may be defined, of dimension $m \times 1$, such that $m = n + p$, that is

$$Z_{k-1} = [X_{k-1}^T, U_{k-1}^T]^T \qquad (5.52)$$

A matrix W can now be introduced, with dimensions $n \times m$, such that

$$W = [\phi \mid \Delta]n \qquad \overset{n \quad p}{}$$

(5.53)

Equation (5.51) then becomes

$$X_k = WZ_{k-1}$$

(5.54)

W is the matrix of actual system parameters, so let W_r be the matrix of estimated parameters based on r sets of measurements.

The best estimate is obtained by minimizing a cost function containing the squared equation errors over the observation interval 1 to r. The error is expressed as

$$e_k = X_k - W_r Z_{k-1}$$

(5.55)

and the cost function is

$$J(W_r) = \sum_{k=1}^{r} (X_k - W_r Z_{k-1})^T (X_k - W_r Z_{k-1})$$

(5.56)

This is minimized with respect to W_r, that is

$$\frac{\partial J(W_r)}{\partial W_r} = 0$$

therefore

$$\sum_{k=1}^{r} X_k Z_{k-1}^T = W_r \sum_{k=1}^{r} Z_{k-1} Z_{k-1}^T$$

(5.57)

Now define a matrix $P_r(m \times m)$ such that

$$P_r^{-1} = \sum_{k=1}^{r} Z_{k-1} Z_{k-1}^T$$

(5.58)

then

$$W_r P_r^{-1} = W_{r-1} P^{-1} + [X_r - W_{r-1} Z_{r-1}] Z_{r-1}^T$$

(5.59)

therefore

$$W_r = W_{r-1} + [X_r - W_{r-1} Z_{r-1}] Z_{r-1}^T P_r$$

(5.60)

Equation (5.60) is the recursive identification algorithm, and the initial setting W_0 may be taken as zero. As it is undesirable to include matrix inversion in the algorithm, the matrix inversion lemma is applied, giving

$$P_r = P_{r-1} - P_{r-1} Z_{r-1} [Z_{r-1}^T P_{r-1} Z_{r-1} + 1]^{-1} Z_{r-1}^T P_{r-1}$$

(5.61)

In this expression, the term in square brackets is a scalar quantity, and the time-variable matrix P_r is a strictly decreasing function of the sample number.

For rapid conversion, the initial value P_0 is taken as a unit matrix, multiplied by a large constant.

5.8.3 Test signals

Pseudo-random binary sequence[36] (PRBS) inputs were employed to perturb the system and generate data for the identification algorithm. The sequences were carefully chosen in relation to the system dynamics, and of small amplitude to ensure linearity. From computer simulation studies, it was found that a seventh-order PRBS is suitable for this application, and a sequence Y_r was used as input to the reference of the exciter. A transformed sequence[36] C_r was applied to the turbine valving, and

$$C_r = (-1)^r Y_r \qquad (5.62)$$

5.8.4 Simulation studies

The identification algorithm was first tested by simulation. The system was modelled by the non-linear equations (5.1) to (5.19) with measured values of parameters, and the PRBS inputs were applied simultaneously to the exciter and prime mover. A fifth-order linear model was identified, and the responses of this model and the non-linear model to various inputs were compared. To further assess the quality of identification, the responses of two linearized analytical models were also computed. These were obtained by reducing the order of the non-linear model to ninth and fifth order, and in each case linearizing the equations by considering small perturbations about the steady-state operating condition at which identification was performed.

The responses are compared in Fig. 5.14. Only three of the five quantities are shown, but the other two show a similar pattern. Both the ninth-order analytical model and the fifth-order identified model provide a reasonable representation of the small-signal dynamics of the non-linear system, but the fifth-order analytical model is considerably less accurate, presumably due to the approximations and assumptions arising in the reduction process. In this model, the damping coefficient has been adjusted to compensate for the omission of damper windings.

It appears, therefore, that identification is particularly useful in obtaining valid low-order models. The value of this observation lies in the fact that it is much easier to design controllers for low-order systems, and subsequent implementation is also facilitated. The simulation studies served to establish that the algorithms are satisfactory, lead to the above conclusions, and set targets against which to assess experimental results.

5.8.5 Experimental identification results

The PRBS signals were generated by simulating a seventh-order shift register in real time in the minicomputer. A complete period of the PRBS lasts 12·7 s, but it was found that equally good results may be obtained using data from the first 8 s only. The measured responses were filtered digitally, using an exponential smoothing algorithm.

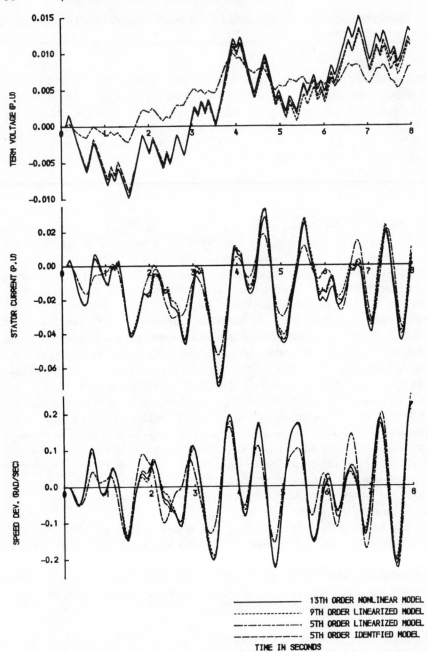

TIME IN SECONDS

———————— 13TH ORDER NONLINEAR MODEL
.... 9TH ORDER LINEARIZED MODEL
—·—·—·—·— 5TH ORDER LINEARIZED MODEL
—————————— 5TH ORDER IDENTFIED MODEL

Fig. 5.14   Simulated responses to PRBS inputs

With the microalternator running at a chosen operating point (defined by the active and reactive power at the machine terminals), the PRBS signals were output via digital-to-analogue converters and added to the constant reference settings. The resulting perturbations of the states were sampled every 20 ms and 400 data sets processed by the identification algorithm. Both the identified models and input sequences were stored, and the response of the model to the same inputs was compared with that of the laboratory system.

A typical set of results is shown in Fig. 5.15, and it appears that the quality of identification is good—although, understandably, not quite up to the standard achieved in simulation.

The amplitude and clock interval used to produce the PRBS signals are variable, and the responses of the identified model and real system were also compared with inputs other than those used to effect identification. This is illustrated in Fig. 5.16, which again shows that the identification is satisfactory. In comparison, the response of the ninth-order linearized analytical model to the same inputs does not follow that of the real system nearly as well. This is also shown in Fig. 5.16.

5.8.6 Controllers

The design procedure is based on the discrete linear model defined by Eqn. (5.51), with parameters obtained by on-line identification. Linear optimal controllers were designed by minimizing the quadratic performance index

$$J = \sum_{k=0}^{Kf-1} \tfrac{1}{2}\{\|X_k\|^2 Q + \|U_k\|^2 R\} \tag{5.63}$$

using the discrete matrix Riccati equation, with the usual constraints on the weighting matrices Q and R.

For example, a controller was designed at the operating point $P_t = 0.8$ p.u., $Q_t = 0$, i.e. the linear model was identified at this point. The diagonal elements of the weighting matrices are

$$Q_{\text{diag}} = 600, 200, 0, 5, 10$$

$$R_{\text{diag}} = 10^5, 10$$

All off-diagonal elements are zero.

The corresponding feedback gains were calculated to be

$$
\begin{array}{ccccc}
V_t & I_t & p\delta & G_V & T_M \\
\begin{bmatrix} -10.9 & -3.00 & 3.44 & 0.61 & 9.62 \\ 0.93 & 0.05 & -0.80 & -0.35 & -2.38 \end{bmatrix} & & & & \begin{matrix} v_{fd} \\ u_g \end{matrix}
\end{array}
$$

The controllers were implemented on the laboratory model turbogenerator, and tested over a wide range of operating conditions, including the application of short circuits on the transmission system, which create large disturbances during which the system is highly non-linear.

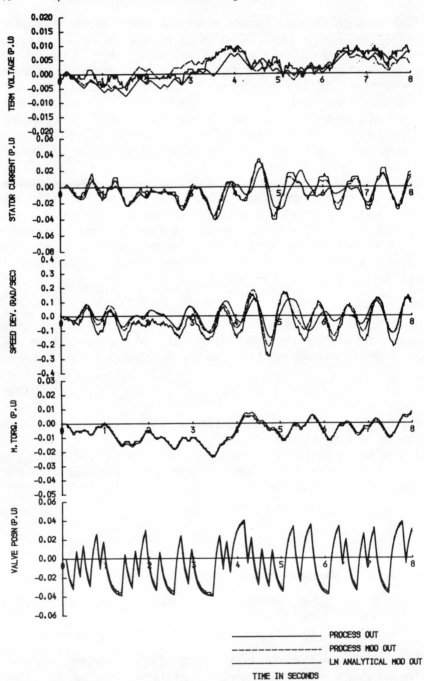

Fig. 5.15   *Comparison of responses*

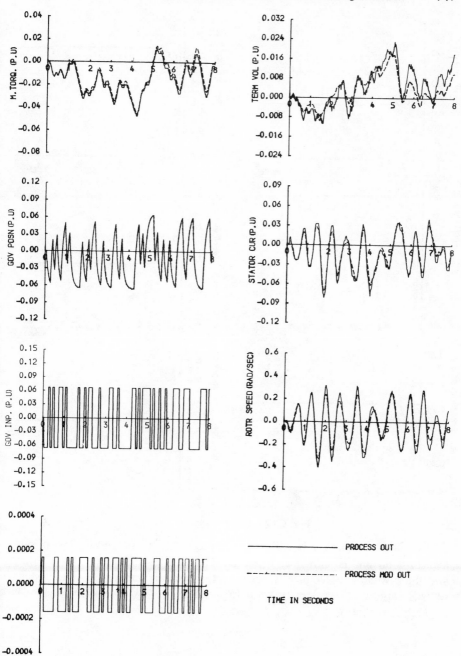

Fig. 5.16 *Experimental identification*

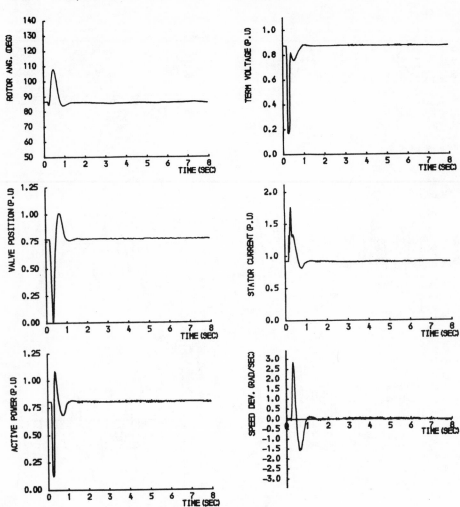

Fig. 5.17   *Effect of short circuit*

The effect of a three-phase short-circuit to earth at the sending end of one transmission line is shown in Fig. 5.17. The fault was cleared after 120 ms, and the postfault impedance is equal to the prefault impedance. The system returns rapidly to the steady state $(P_t = 0.825$ p.u., $Q_t = -0.005$ p.u.) with negligible oscillation.

5.9 Multimachine systems

5.9.1 Introduction

The controllers developed in this chapter, and most of those described in the literature, have been designed and tested on a single machine. However, it is necessary to assess the performance of the controllers in a multimachine environment in which there are interactions between the various machines. Also, if multivariable controllers are introduced on new installations, they will be operating in parallel with a large number of existing conventional controllers, most of which are slow acting. Some of the questions which arise are:

1. Will controllers which have been designed for a single unit operate satisfac- torily in a multimachine system?
2. Even if they will, are these the best controllers?
3. Is it necessary to provide information other than that locally available? (such as the 'other inputs' indicated in Fig. 5.1.)
4. What is the effect of interactions between machines when some have multivariable controllers and others have conventional controllers?

A multimachine simulation program was developed to assess these and other factors, and the model is described briefly in this section.

It was decided to use a detailed representation of each boiler-turbogenerator unit, and consequently the digital simulation involves the simultaneous solution of the complete non-linear model, together with the solution of the linear voltage and current equations of the network. Load non-linearities and load effects were neglected. Computer programs using detailed machine models may employ iterative techniques to couple the network with the machines,[37] or arrange the network and generator equations in a manner suitable for a closed-form derivative computation algorithm.[38] The method used here was initially based on these techniques, but all machines are represented in detail [Eqns. (5.1) to (5.19)] with the complete boiler-turbine (Fig. 5.3), thus giving a fifteenth-order model. It was found that neither of the above approaches is satisfactory with such a complicated representation.

5.9.2 Modelling

Each generator is described by the non-linear equations expressed in Park's reference frame, which is fixed to the machine rotor. The network is described by lumped impedances, and the solution of currents and voltages at each specified node (to which either a synchronous machine or a load is connected) is with respect to a common reference frame rotating at synchronous speed. During disturbances, the speeds of machines change, and therefore their individual reference frames oscillate with respect to the common reference frame.

The derivation of the equations is long and complex,[37, 38] and will not be reproduced here. The load-flow programme employs Gauss' iterative method, and the coupling of network and machine equations is achieved by means of the auxiliary prediction formula.

The programme reads the initial nodal specifications and generates the steady-state load-flow solution. The initial conditions for the machines are then calculated, and the non-linear equations are solved using a numerical method based on the Runge-Kutta technique. Provision is made for various control arrangements on the machines. The iterative approach to solving the network was found to be slow, and was replaced by a method based on matrix manipulation.

The step length used for integration is an important factor in the solution of these equations, as it greatly influences both the execution time of the programme and the accuracy of the results. In order to obtain accurate solutions of the detailed models used here, the step length had to be reduced to 0·0001 s.

5.9.3 Two-machine system

As an example, a two-machine system is considered. This is shown in Fig. 5.18. Machine A is connected to an infinite busbar through a transformer and two parallel transmission lines, and this arrangement is identical to the single-machine system considered previously. Machine B may be taken as adjacent to, or remote from, machine A, depending on the impedance between busbars 2 and 3.

The machines, transformers and two parallel transmission lines are given the parameter values listed in Appendix A, and the impedance between bus-

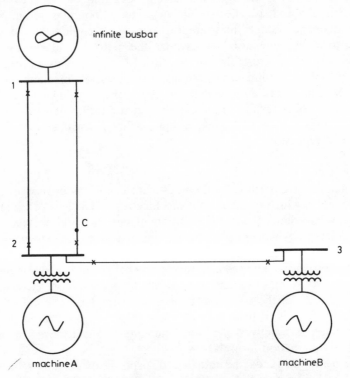

infinite busbar

machine A   machine B

Fig. 5.18   *Two-machine system*
x = circuit breaker

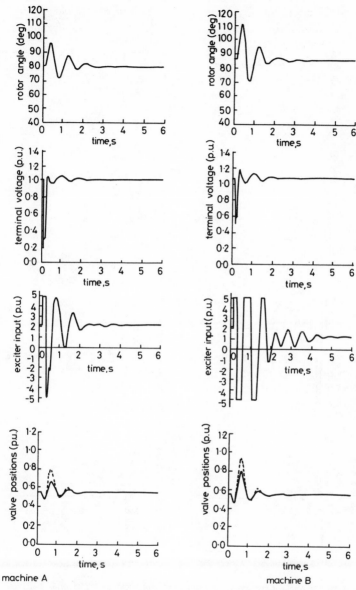

machine A   machine B

Fig. 5.19 *Transient response of multi-machine system—conventional controllers*

bars 2 and 3 is $0.025 + j0.35$ p.u. The response to a three-phase short circuit to earth at point C has been computed, and is shown in Figs. 5.19 and 5.20. The duration of the fault is 120 ms.

Firstly (Fig. 5.19), both machines were assumed to have conventional controllers, with thyristor exciters and electrohydraulic governors. Excitation control is provided by automatic voltage regulators with speed stabilizers.[16]

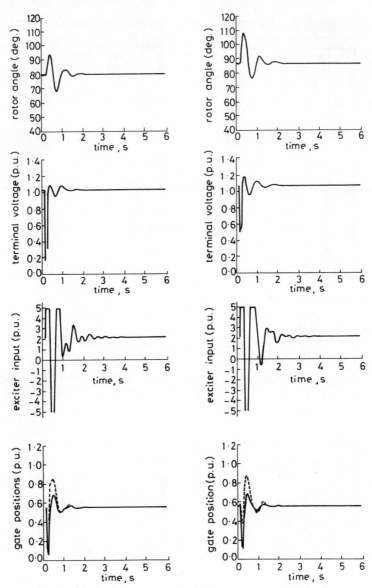

Fig. 5.20   *Transient response of multi-machine system—multivariable controllers*

Subsequently, the governors and a.v.r.'s on both machines were replaced by the multivariable controllers described in Section 5.5.3. The response of this system with controller 1 is shown in Fig. 5.20.

In general, relatively simple machine models are used in analysing or simulating the performance of large multimachine systems, and the controllers are usually omitted. Detailed models are necessary to assess the performance of control systems, but the representation of a small number of machines is sufficient for this purpose. The system shown in Fig. 5.18 can easily be made into a three-machine system by replacing the infinite bus by another generator.

5.10 Acknowledgements

The author is grateful to the Science Research Council (U.K.) for financing the laboratory model turbogenerator, and to the University of Liverpool for laboratory facilities and technical support. He wishes to thank Dr. R. T. Pullman, Mr. S. M. Osheba, and Mr. M. M. Sharaf for their substantial contributions to this project.

5.11 Appendix A

System parameters

Microalternator: (values, per unit)

$H = 3\cdot25$, $R_{fd} = 0\cdot0015$, $R_{kd} = 0\cdot0078$, $R_{kq} = 0\cdot0084$, $R_m = 0\cdot005$, $X_{ad} = 1\cdot86$, $X_{aq} = 1\cdot77$, $X_{fd} = 1\cdot97$, $X_d = 2\cdot0$, $X_q = 1\cdot91$, $X_{kd} = 1\cdot94$, $X_{kq} = 1\cdot96$, $K_d = 0$.

Exciter:

$\tau = 0\cdot01$ s, Voltage limits ±5 p.u.

Turbine and governors:

$K_H = 0\cdot24$, $K_I = 0\cdot34$, $K_L = 0\cdot42$, $\tau_H = 0\cdot30$ s, $\tau_I = 0\cdot30$ s, $\tau_L = 0\cdot72$ s, $\tau_R = 10\cdot0$ s, $\tau_G = 0\cdot1$ s, $\tau_C = 0\cdot1$ s, $\tau_p = 0\cdot2$ s, $\tau_b = 6\cdot0$ s, $\tau_f = 8\cdot0$ s, $\tau_1 = 4\cdot5$ s, $\tau_2 = 27$ s, $\tau_3 = 3\cdot49$ s, $K_1 = 4\cdot5$, $K_2 = 18\cdot22$, $K_3 = 1\cdot32$, $P_r = 1\cdot2$, $\tau_T = 0\cdot5$ s.

Maximum opening and closing rates for both intercept and inlet valves are $\pm6\cdot7$ p.u.
Position limits 0–1·0 p.u.

Transmission system:

Transformer leakage reactance $= 0\cdot1$ p.u.
Two parallel lines, each consisting of two series RLC π-sections, with $R = 0\cdot025$ p.u., $X = 0\cdot35$ p.u., $B = 5 \times 10^{-4}$ p.u.

5.12 References

1 ALDRED, A. S.: 'Electronic analogue computer simulation of multimachine power systems networks', *Proc. IEE*, 1962, **109**, pp. 195–202
2 SHACKSHAFT, G.: 'General purpose turboalternator model', *Proc. IEE*, 1963, **110**, pp. 703–713
3 NICHOLSON, H.: 'Dynamic optimisation of a boiler-turboalternator model', *Proc. IEE*, 1966, **113**(2), pp. 385–399
4 KUNDUR, P., and BAYNE, J. P.: 'A study of early valve actuation using detailed prime mover and power system simulation', *IEEE Trans.*, 1975, **PAS-94**(4), p. 1275
5 VENIKOV, V. A.: 'Theory of similarity and simulation' (Macdonald, London, 1969)
6 NEWTON, M. E., and HOGG, B. W.: 'Optimal control of a microalternator', *IEEE Trans.*, 1976, **PAS-95**, pp. 1821–33
7 PULLMAN, R. T., and HOGG, B. W.: 'Discrete state-space controller for a turbogenerator', *Proc. IEE*, 1979, **126**, pp. 87–92
8 AHSON, S. I., HOGG, B. W., and PULLMAN, R. T.: 'Integrated control system for a turbogenerator designed by inverse Nyquist array method', *IEEE Trans.*, 1979, **PAS-98**, pp. 543–553
9 GRAUPE, D.: 'Identification of systems' (Van Nostrand-Reinhold Company, 1972)
10 HOPE, G. S., NICHOLS, S. T., and CARR, J.: 'Measurement of transfer functions of power system components under operating conditions', *IEEE Trans.*, 1977, **PAS-96**, pp. 1798–1808
11 SHARAF, M. M., and HOGG, B. W.: 'Identification and control of a laboratory model turbogenerator', *Int. J. Cont.*, 1980, **31**, pp. 723–739
12 SANDOZ, D. J., and WONG, O.: 'Design of hierarchical computer control system for industrial plant, Pt. I', *Proc. IEE*, 1978, **125**, pp. 1290–1298
13 ABDALLA, O. H., and WALKER, P. A. W.: 'Optimal control of a laboratory power system model with output feedback', IEEE Summer Power Meeting, 1979, Paper No. A79 451–6
14 JOHNSTONE, L. R., and MARSLAND, C. R.: 'Distributed computer control for power station plant', *Electron. & Power*, May, 1978, pp. 374–378
15 FENWICK, D. R., and WRIGHT, W. G.: 'Review in excitation systems and possible future development', *Proc. IEE*, 1976, **123**, pp. 413–420
16 WATSON, W., and MANCHUR, G.: 'Experience with supplementary damping signals for generator static excitation', *IEEE Trans.*, 1973, **PAS-92**, pp. 199–203
17 IEEE COMMITTEE REPORT: 'Dynamic models for steam and hydro turbines in power system studies', *IEEE Trans.*, 1973, **PAS-92**, pp. 1904–1915
18 HERNANDEZ, R., and FRERIS, L. L.: 'Micro-machine models for steam power plant', *Proc. IEE*, 1974, **121**, pp. 491–499
19 HAM, P. A. L., JENKINS, K., and MIKHAIL, S. E.: 'Performance and control capabilities of electro-hydraulic governing systems for steam turbogenerator units', *Fifth Iranian Conference on Electrical Engineering*, 1975, Shiraz, Iran
20 LAUBLI, F., and FENTON, F. H.: 'The flexibility of a super-critical boiler as a partner in power system design and operation', *IEEE Trans.*, 1971, **PAS-90**, pp. 1719–1733
21 PULLMAN, R. T.: 'Co-ordinated excitation and governor control of turbogenerators using state-space techniques', 1977, Ph.D. Thesis, University of Liverpool
22 GROSS, G., and BERGEN, A. R.: 'A class of new multi-step integration algorithms for the computation of power system dynamical response, *IEEE Trans.*, 1977, **PAS-96**, p. 293
23 GEAR, C. W.: 'The automatic integration of stiff ordinary differential equations', *Proc. IFIPS Congress*, 1969, North Holland, Amsterdam, p. 187
24 ANDERSON, J. H., HUTCHISON, M. A., WILSON, W. J., ZOHDY, M. A., and APLEVICH, J. D.: 'Micro-alternator experiments to verify the physical realisability of si-

mulated optimal controllers, and associated sensitivity studies', *IEEE Trans.*, 1978, **PAS-97**, pp. 649-658

25 BURROWS, P. J., and DANIELS, A. R.: 'Digital excitation control of a.c. turbogenerators using a dedicated microprocessor', *Proc. IEE*, 1978, **125**(3), pp. 237-40

26 MOYA, O. E. O., and CORY, B. J.: 'Online control of generator transient stability by minicomputer', *Proc. IEE*, 1977, **124**(3), pp. 252-58

27 OSHEBA, S. M., and HOGG, B. W.: 'Multi-variable controller for a turbogenerator', IEEE Winter Power Meeting, 1980, Paper No. A80 106-5

28 DANIELS, A. R., DAVIS, D. H., and PAL, M. K.: 'Linear and non-linear optimisation of power system performance', *IEEE Trans.*, 1975, **PAS-94**, pp. 810-18

29 SMITH, H. W., and DAVISON, E. J.: 'Design of industrial regulators', *Proc. IEE*, 1972, **119**(8), pp. 1210-16

30 PORTER, B.: 'Optimal control of multivariable linear systems incorporating integral feedback', *Electron. Lett.*, 1971, pp. 170-172

31 DeMELLO, F. P., HANNETT, L. N., and UNDRILL, J. M.: *IEEE Trans.*, 1978, **PAS-97**, pp. 1515-1522

32 PULLMAN, R. T., and HOGG, B. W.: 'Instrumentation for testing and control of laboratory model turbogenerator', *IEEE Trans.*, 1978, **IM-27**, pp. 188-192

33 ANDERSON, J. H., HUTCHINSON, M. A., WILSON, W. G., RAINA, V. M., OKONGWU, E., and QUINTANA, V. H.: 'A practical application of optimal control using a microalternator', *IEEE Trans.*, 1977, **PAS-96**, pp. 729-740

34 OSHEBA, S. M., and HOGG, B. W.: 'Compensation of rotor backswing in a microalternator', *Indian Journal of Technical Education*, 1980

35 SINHA, N. K., and SEN, A.: 'Critical evaluation of on-line identification methods', *Proc. IEE*, 1975, **110**, pp. 1153-58

36 BRIGGS, P. A. N., and GODFREY, K. R.: 'Pseudo-random signals for the dynamic analysis of multi-variable systems', *Proc. IEE*, 1966, **113**, pp. 1259-67

37 PRABHASHAUKAR, K., and JANISCHEWSKYJ, W.: 'Digital simulation of multi-machine power systems for stability studies', *IEEE Trans.*, 1968, **PAS-87**, pp. 73-80

38 UNDRILL, J. M.: 'Structure in the computation of power-system non-linear dynamical response', *IEEE Trans.*, 1969, **PAS-88**, pp. 1-6

Chapter 6

Modelling of gas supply systems

M. H. Goldwater and A. E. Fincham

List of principal symbols

a	isothermal speed of sound, $a^2 = (\partial p/\partial \rho)_T$
A	cross sectional area of pipe
c	isentropic speed of sound, $c^2 = (\partial p/\partial \rho)_S$
C^+, C^-	labels for characteristics
d_i	demand at node i
D	internal pipe diameter
e	specific internal energy
E	pipe efficiency
f	Fanning friction factor
$f(p, q, T)$	expression for $(\partial p/\partial t)$ from conservation equations, $\partial p/\partial t = f(p, q, T)$
f_i	value of f at mesh point in finite difference approximation
f_{FT}, f_{PT}, f_{SPL}	friction factors for fully turbulent flow, partially turbulent flow, and smooth pipe law
$F(\partial p/\partial t, p, q, T)$	expression for finite difference form of mass conservation equation
F_f	drag factor for partially turbulent flow
g	acceleration due to gravity
$g(p, q, t)$	expression for $\partial q/\partial t$ from conservation equation, $\partial q/\partial t = g(p, q, T)$
$G(\partial q/\partial t, p, q, T)$	expression for finite difference form of momentum equation
h	specific enthalpy, $h = e + p/\rho$; also finite difference (element) length increment
h_j	finite difference (length) increment for jth pipe

$h(p, q, t)$ expression for $\partial T/\partial t$ from conservation equations, $\partial T/\partial t = h(p, q, T)$

$\mathbf{H}(\partial \mathbf{T}/\partial t, \mathbf{p}, \mathbf{q}, \mathbf{T})$ expression for finite difference form of energy equation

k, k_e roughness, effective roughness of pipe

l_j length of jth pipe

L_k set of pipes for which kth node is left node

m Mach number, $m = v/a$

M molecular weight of gas

p pressure

p_i pressure at ith mesh point in finite difference (element) mesh; pressure at ith node of network

$\mathbf{p}$ vector of pressures at mesh points in finite difference (element) mesh

$\hat{p}$ approximate p used in finite element method

p_O pressure at compressor outlet

p_I pressure at compressor inlet

p_{set} set point pressure for compressor outlet pressure

q flow

q_i flow at ith mesh point in finite difference (element) mesh

$\mathbf{q}$ vector of flows at mesh points in finite difference (element) mesh

$\hat{q}$ approximate q used in finite element method

r_{set} set point pressure ratio for compressor, $p_O/p_I = r_{set}$

R gas constant ($8{\cdot}3143$ J/kmol)

Re Reynolds number, $\mathrm{Re} = \rho v D/\mu$

R_k set of pipes for which kth node is right node

S_k set of pipes incident on the kth node

t time

T temperature

$\mathbf{T}$ vector of temperatures at mesh points in finite difference (element) mesh

$\hat{T}$ approximate T used in finite element method

u_i chapeau function for ith mesh point in finite element method

v velocity

$\bar{V}$ bulk mean velocity

w weighting function in finite element method

$\{w\}$ set of weighting functions in finite element method

W_{max} maximum power of compressor

x distance along pipe

x^+, x^- distance along characteristics C^+, C^-

Z	compressibility
α	dimensionless parameter, $\alpha = \dfrac{1}{A} \displaystyle\iint_A \dfrac{\rho v^2}{\bar{\rho}\bar{V}^2}\, dA$
Δt	time step
Δx	mesh size in finite difference (element) method
θ	angle of inclination of pipe to the horizontal
κ	polytropic index for gas compression
λ	multiplier in MOC equations
μ	gas viscosity
ρ	density of gas
$\bar{\rho}$	bulk mean density
τ	tangential stress at pipe wall
Ω	heat flow from ground to pipe for 1 metre length of pipe

6.1 Introduction

The modelling of gas supply systems has been given much impetus from the major changes that have occurred in the gas industry in the last two decades. In this time we have changed from manufactured gas using coal and more recently oil feedstocks to natural gas. This change has necessitated the construction of extensive transmission systems to bring the natural gas from the often remote locations where it is discovered and fed into existing transmission and distribution systems. The design and operation of these new transmission systems has provided some important applications for computer modelling and the consequent financial benefits that help to justify the not insignificant costs of development and continuing support of the computer programs.

An example of the complexity of the transmission systems that must be modelled is shown in Fig. 6.1. This is a map of the National Gas Transmission System in Great Britain and shows the routes of pipelines and the location of compressor stations and storage installations. This system is designed and operated by British Gas: the gas is received into the network at the shore terminals and is transmitted to the regional transmission and distribution systems. The pipelines are mainly 36-in diameter steel pipe operating at pressures between 35 and 70 bar, the compressors are the centrifugal type driven by gas turbine engines fuelled by natural gas. Liquid natural gas (LNG) is stored at strategic locations, it is evaporated and fed into the system at times of peak demand when pipeline and compressors are loaded to full capacity. Gas is also stored in salt cavities leached out underground, but this is limited to a few locations such as East Yorkshire where the salt strata occur at sufficient depth.

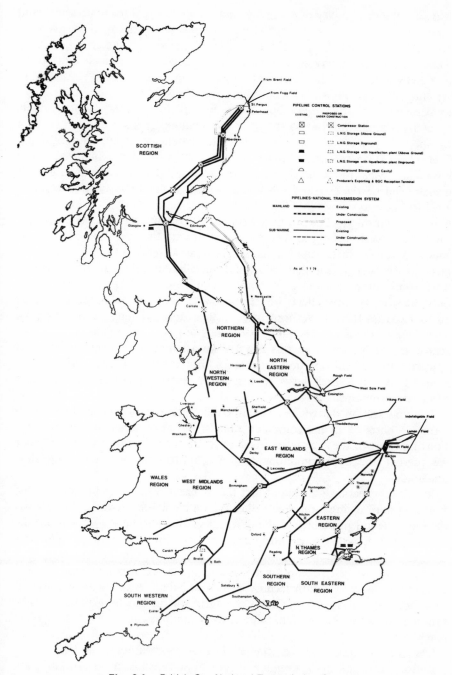

Fig. 6.1   *British Gas National Transmission System*

Before the availability of computer models all calculations of pressures and flows in transmission and distribution systems were based on steady-state flow and were limited to relatively simple networks. The first computer models were also steady state but were capable of solving much more complex networks. With the increased emphasis on high-pressure transmission systems it was realized that steady-state models were not sufficient; in high-pressure networks dynamic effects are important. Firstly there is the dynamic effect associated with the gas stored in the pipeline. This stored gas is known as linepack and it can be exploited to help balance supply and demand: linepack is depleted when demand is high and replenished when demand is low. This cycle will usually take place over a 24-hour period with high demand during the day and low demand at night, and the dynamic changes are fairly slow as demands will generally change significantly in a time scale of hours. In contrast to this, we have very rapid dynamic effects associated with sudden changes on the grid such as can occur with the operation of compressors and valves; these rapid dynamics have time scales of seconds and can give rise to such phenomena as rapidly oscillating values of pressure and flow. Some computer models of dynamic behaviour have attempted to model only the slow dynamic effects whilst others attempt to model both. In the past 10 to 15 years there have been more than a dozen different computer programs for modelling the dynamic behaviour of gas transmission systems presented in the literature. In this chapter we attempt to explain the different approximations and solution techniques that have been used so that the reader may have some basis to assess the relative merits of the different computer programs.

The first section derives the fundamental equations of pipeline flow which form the basis of all the computer models. We also discuss the approximations inherent in these equations and examine their limitations. In particular we present the approximations that are appropriate for modelling slow dynamics.

In the next section we present first the different solution techniques for the simple case of a single pipeline rather than a complete network; this enables the fundamental principles of the different methods to be shown without the unnecessary complications introduced in modelling the network. We then discuss the computer models that have been presented in the literature, classifying them by their solution methods and the approximations used. Finally we show how the solution techniques can be adapted for modelling a complete network by introducing extra equations for the conditions that must be satisfied at network junctions and across compressors and other pipeline machinery.

The final section discusses the application of computer models to real problems that arise in the design and control of transmission networks. These problems do not arise in a form that is ready-made for solution by using a simulation model, so that often considerable skill must be employed to convert the problem into a question or series of questions that can be answered by the model, and then these answers must again be translated back to the terms of the real problem. This procedure is discussed briefly and literature references are given to papers which discuss applications in more detail. The techniques of short-term prediction by simulation models for control of gas networks are being developed and the problems that have to be overcome are described.

We do not attempt to compare the effectiveness of the different models mentioned; this is because very little work has been done on comparing models and there is insufficient objective evidence on the relative merits of different solution methods in practical application. Probably the most important factor in determining effectiveness is the design and coding of the computer program that represents the model; the same method can have good or bad implementations.

6.2 Pipeline flow equations

The flow of gas in a pipeline is turbulent and a rigorous development of the equations of flow requires the application of the methods used in turbulent boundary-layer theory as described, for example, in Shapiro.[32] Since this would be lengthy and unnecessarily complicated we shall follow the general practice and assume that the flow is one-dimensional; the justification for this approximation is discussed below. We also make the assumption that the pipe is straight with constant, circular cross section. We can then derive the partial differential equations which describe the flow by applying the laws of conservation of mass, momentum and energy to an infinitesimal control volume.

Mass. Applying conservation of mass to the control volume shown in Fig. 6.2(a)

$$\frac{\partial}{\partial t}(\rho A) + \frac{\partial}{\partial x}(\rho v A) = 0 \qquad (6.1)$$

Momentum. Applying conservation of momentum to the control volume shown in Fig. 6.2(b)

$$\frac{\partial}{\partial t}(\rho v A) + \frac{\partial}{\partial x}(pA + \rho v^2 A) + |\tau|\pi D + \rho A g \sin \theta = 0 \qquad (6.2)$$

Energy. Applying conservation of energy to the control volume shown in Fig. 6.2(c)

$$\frac{\partial}{\partial t}[(e + \tfrac{1}{2}v^2)\rho A] + \frac{\partial}{\partial x}[(h + \tfrac{1}{2}v^2)\rho v A] - \Omega + \rho A g v \sin \theta = 0 \qquad (6.3)$$

We can regard these equations for four unknowns p, ρ, v, T; the values of other quantities being known or calculable from these four variables. Since we

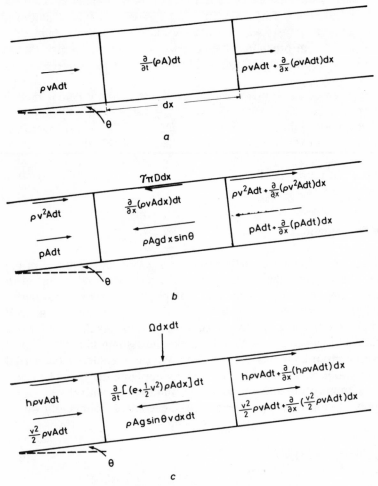

Fig. 6.2 *(a) Control volume for mass conservation; (b) Control volume for momentum conservation; (c) Control volume for energy conservation*

have only three equations for four unknowns we need another equation to solve the system; this is provided by the equation of state for the gas

$$p = \rho \frac{ZRT}{M} \qquad (6.4)$$

In this equation Z is a variable compressibility so the gas is treated as non-ideal.

6.2.1 Friction factor

In fully developed steady flow the shear stress can be represented by the Fanning friction factor f

$$f = \frac{|\tau|}{\frac{1}{2}\rho v^2} \tag{6.5}$$

f is a dimensionless quantity and in general is a function of the Reynolds number and the roughness of the pipewall

$$f \equiv f\left(\text{Re}, \frac{k}{D}\right) \tag{6.6}$$

For high-speed compressible flow there is some influence of Mach number (v/c) on f but this effect is negligible for gas transmission $(v/c \sim 0.02)$.

A detailed account of the theoretical background and experimental data for friction factors in gas transmission is given by Uhl et al.[38] They recommend the following equations:

$$\sqrt{\frac{1}{f_{SPL}}} = 4 \log\left(\frac{\text{Re}}{\sqrt{1/f_{SPL}}}\right) - 0.6 \tag{6.7a}$$

$$\sqrt{\frac{1}{f_{PT}}} = F_f \sqrt{\frac{1}{f_{SPL}}} \tag{6.7b}$$

$$\sqrt{\frac{1}{f_{FT}}} = 4 \log\left(\frac{3.7D}{k_e}\right) \tag{6.8}$$

The transition from partially turbulent flow [Eqns. (6.7a) and 6.7b)] to fully turbulent flow [Eqn. (6.8)] takes place at a critical Reynolds number where $\sqrt{1/f_{PT}}$ and $\sqrt{1/f_{FT}}$ are equal. The drag factor F_f and effective roughness are adjustable parameters which can be chosen to allow for the effects of bends and fittings in a real pipeline.

Another popular equation for transmission is the Panhandle 'A' formula

$$\sqrt{\frac{1}{f}} = 6.78 \, \text{Re}^{0.073} \, E \tag{6.9}$$

The efficiency E is analogous to the drag factor in Eqn. (6.7b) and the Panhandle 'A' friction factor is very similar to the partially turbulent friction factor of Eqns. (6.7a) and (6.7b); Eqn. (6.9) has the advantage that it can be solved for $\sqrt{1/f}$ explicitly: Eqn. (6.7a) requires iteration.

The use of a friction factor to represent the wall shear stress is only justified theoretically for steady flow, but it is common practice to use the same friction factor equations for unsteady flow. It is reasonable to expect this to be valid for small perturbations around a steady flow condition and some experimental evidence for this is given in a paper by Brown et al.[7] For large, rapid flow changes we cannot expect the approximation to be very good, but it is often applied to these conditions. There is a need for more work in this area to establish the limitations of the use of steady-flow friction factors.

6.2.2 Heat conduction

To solve Eqns. (6.1) to (6.4) we need to know or be able to calculate the value of the heat flow term Ω. There are two important special cases:

1. Isothermal flow, $T = $ constant; the energy equation then becomes redundant except to calculate the value of Ω.
2. Adiabatic flow, $\Omega = 0$, including the particular case of isentropic flow.

These two cases may be considered as two extremes. Isothermal flow corresponds to slow dynamic changes where temperature changes within the gas are sufficiently slow to be cancelled out by heat conduction between the pipe and the surrounding soil. Adiabatic flow corresponds to rapid dynamic changes in the gas where the slow-acting heat conduction effects can be neglected. Isentropic flow is only possible when friction effects are negligible, and this is not normally valid in gas transmission problems.

For the general case, where $\Omega \neq 0$ and there is no thermal equilibrium between the pipe and the soil, we need further equations to model the heat conduction process. We do not know of any such model for non-steady flow but Buthod et al.[8] have described a model including heat conduction for steady flow in a gas pipeline.

6.2.3 Validity of one-dimensional approximation

It has been shown that the one-dimensional flow model enables us to derive the basic operations of pipeline flow in a simple fashion. The flow cannot be one-dimensional because viscous effects will produce a velocity profile across the pipe with the local velocity zero at the pipe wall and reaching a maximum in the centre; moreover the flow is turbulent so that there are random motions superimposed upon the mean flow. The departure from the one-dimensional-flow model will be even more pronounced where there are bends and fittings in the pipeline. A more rigorous approach is to apply the integrated equation method as used in turbulent boundary-layer theory. Ward Smith[39] uses this approach for steady flow in pipelines, he then defines a suitable set of mean-flow properties at a cross section and expresses the integrated equations in terms of these variables. The equations in the mean-flow properties are similar to the one-dimensional equations for steady flow except that there are variable coefficients multiplying some of the terms and there are extra terms introduced involving the derivatives of these coefficients. For turbulent flow, Ward Smith argues that these coefficients are always close to unity, and simplifies his equations by replacing them by unity; this then gives equations which are exactly the same as the one-dimensional equations for steady flow. As an illustration of this process we shall consider the coefficient α defined by Ward Smith

$$\alpha = \frac{1}{A} \iint\limits_{A} \frac{\rho v^2}{\bar{\rho} \bar{V}^2} \, dA \tag{6.10}$$

where

$$\bar{V} = \frac{1}{A} \iint\limits_{A} v \, dA \tag{6.11}$$

and

$$\bar{\rho} = \frac{\iint\limits_{A} \rho v \, dA}{\iint\limits_{A} v \, dA} \tag{6.12}$$

The term $(\partial/\partial x)\rho v^2$ in Eqn. (6.2) is replaced by

$$\alpha \frac{\partial}{\partial x} \bar{\rho} \bar{V}^2 + \bar{\rho} \bar{V}^2 \frac{\partial \alpha}{\partial x}$$

which is equivalent if $\alpha = 1$. Ward Smith quotes values of α between 1 and 1.02 for turbulent flow. This is supported by Brown et al.[7] who quote the following values for fully developed steady flow in a pipeline:

Re	α
Laminar flow	4/3
2500	1·113
10^4	1·049
10^5	1·020
10^6	1·012
10^7	1·008
∞	1

For gas transmission we are generally in the region of Re $\sim 10^7$ so the approximation $\alpha = 1$ is very good.

In the regions of influence on flow of bends and fittings there will be larger departures from unity though this will be highly localized. In non-steady flow we must also expect larger discrepancies from the one-dimensional flow approximation particularly when there are large, rapid changes in conditions.

6.2.4 Simplification of equations
We have already introduced one important simplification of Eqns. (6.1) to (6.3) for slow dynamic effects: that is the assumption of isothermal flow. Even

with this assumption we must still solve a pair of non-linear hyperbolic partial differential equations. If we neglect the friction term we get the equations of classical inviscid gas dynamics in which pressure waves propagate through the gas at the speed of sound without any damping. If instead we neglect inertia terms we get creeping motion which is described by parabolic partial differential equations akin to heat conduction and diffusion, and the phenomenon of wave motion disappears. Real pipeline flow is a mixture of these two extremes but there is the possibility of neglecting terms to make the computation much easier.

To decide which terms may be neglected we need magnitude estimates for each term for characteristic values of the variables. As an example we shall consider a high-pressure transmission line where the dynamic variations are determined by the fluctuations in demand with significant changes occurring on the time scale of hours. The characteristic values for the variables for these conditions may be taken as follows:

$$v = 10 \text{ m/s} \qquad D = 1 \text{ m} \qquad \rho = 50 \text{ kg/m}^3$$

$$f = 0.002 \qquad t = 1 \text{ hour} = 3600 \text{ s} \qquad x = 100 \text{ km} = 10^5 \text{ m}$$

$$p = 70 \text{ bars} = 70 \times 10^5 \text{ N/m}^2$$

The magnitude estimates for terms in the momentum equation are

$$\frac{\partial}{\partial t}(\rho v A) \approx 0.1 \qquad \frac{\partial}{\partial x}(pA) \approx 50$$

$$\frac{\partial}{\partial x}(\rho v^2 A) \approx 0.04 \qquad f \frac{\rho v^2 \pi D}{2} \approx 15$$

$$\rho A g \sin \theta \approx 385 \sin \theta$$

For these conditions it is reasonable to neglect the inertia terms $\partial/\partial t(\rho v A)$ and $\partial/\partial x(\rho v^2 A)$ as they contribute less than 1% of the friction term. This is the creeping motion approximation referred to above.

We can get more rapid disturbances on a transmission system, for example, a valve opening, a compressor shut-down, a sudden change in demand. Under these conditions the values of the inertia terms in the region of the disturbances will be significant, and creeping motion is not likely to be a satisfactory approximation in that region though it may still be a reasonable approximation to the gross behaviour of the whole system.

It is difficult to decide which terms to neglect and it will depend on how much emphasis is placed on the representation of the more rapid dynamic disturbances. The majority of workers neglect the term $\partial/\partial x(\rho v^2 A)$ but retain $\partial/\partial t(\rho v A)$ though in some cases this decision is influenced by the convenience of this approximation in applying the method of characteristics (see below). Weimann[40] has investigated the most suitable form of approximation by studying the frequency response of a typical transmission pipeline; he concludes that the term $\partial/\partial x(\rho v^2 A)$ can be neglected but that $\partial/\partial t(\rho v A)$ should be retained if disturbances with a cycle time of less than 8 min are important. A strong argument for the retention of both inertia terms is put forward by Rachford[26] and typical problems where inertia is important are presented in his paper and also in Rachford and Dupont.[28]

6.2.5 Mass-flow formulation
Consumer demand for gas is expressed in terms of mass flow so it is more convenient to replace the velocity v in Eqns. (6.1) to (6.3) by the molar flow q defined by

$$q = \frac{\rho v A}{M} \tag{6.13}$$

We shall assume isothermal flow which is the most common approximation. On substituting Eqns. (6.5) and (6.13) into Eqns. (6.1) and (6.2) we obtain the equations

$$\frac{A}{a^2}\frac{\partial p}{\partial t} + M\frac{\partial q}{\partial x} = 0 \tag{6.14}$$

$$\frac{\partial p}{\partial x} + \frac{M}{A}\frac{\partial q}{\partial t} + \frac{M^2}{A^2}\frac{\partial}{\partial x}\left(\frac{q^2}{\rho}\right) + \frac{M^2}{A^2}\frac{2fq|q|}{\rho D} + \rho Ag\sin\theta = 0 \tag{6.15}$$

where

$$a^2 = \left(\frac{\partial p}{\partial \rho}\right)_T \tag{6.16}$$

a is the isothermal speed of sound in the gas.

The friction term contains the expression $q|q|$ instead of q^2, this ensures that the friction force always opposes the motion of the gas.

If we neglect inertia terms then we obtain the simpler form of Eqn. (6.15) for creeping motion

$$\frac{\partial p}{\partial x} + \frac{M^2}{A^2}\frac{2fq|q|}{\rho D} + \rho Ag\sin\theta = 0 \tag{6.17}$$

6.3 Models of gas flow in pipelines

The equations described in Section 6.2 can, with some difficulty, be solved numerically in their entirety. However, the full equations require a large volume of data about the system which is difficult to gather and keep up to date. It is fortunate that for many purposes the equations and the data required to solve them may be simplified. Then everyday pipe-network models of systems run by, for example, gas distribution companies may be constructed using ordinary engineering records and the load data inferred from the customer billing files. In this chapter we first deal with the method of turning the partial differential equations into sets of algebraic equations which can be solved on a computer, and then we give a review of the simplifications that various authors have made. The different levels of simplification are aimed at coping with different extremes of data for the models: a model to be used for planning a transmission system which will be operated steadily can neglect most of the terms which are needed to model a pipebreak.

6.3.1 Numerical analysis for pipelines

An analytic solution can be obtained from the pipeline equations by dropping the friction terms and considering sufficiently simple initial conditions—the equations become the classical equations of gas dynamics. However, for general modelling purposes a numerical model must be used.

The pressure-velocity-energy continuum is evaluated at only a finite number of points along the pipe and intervening values are obtained by interpolating between those points. The form of the interpolation will depend on the technique used to solve the partial differential equations. This will lead to a set of ordinary differential equations in the time variable and there is, of course, a choice of techniques for solving them.

6.3.1.1 Finite differences. Pressures, flows and temperatures are considered at only a finite number of grid points, and the various differentials are constructed in the usual way. If the grid points are at x_i where $x_i = ih$ and $h = l/n$ and $p_i = p(x_i)$, $q_i = q(x_i)$ and $T_i = T(x_i)$, then we can approximate each of the equations (6.14) and (6.15) at or near the midway position between the grid points. We can approximate terms like $\partial p/\partial x(\frac{1}{2}(x_i + x_{i+1}))$ by $(p_{i+1} - p_i)/h$. Making such approximations we derive the following equations (Fincham and Goldwater[13])

$$\frac{1}{2}\frac{A}{a^2}\left(\frac{dp_i}{dt} + \frac{dp_{i+1}}{dt}\right) + M\frac{q_{i+1} - q_i}{h} = 0 \tag{6.18}$$

$$\frac{1}{2}\frac{M}{A}\left(\frac{dq_i}{dt} + \frac{dq_{i+1}}{dt}\right) + \frac{p_{i+1} - p_i}{h} + \frac{M^2}{A^2}\left(\frac{q_{i+1}^2}{\rho_{i+1}} - \frac{q_i^2}{\rho_i}\right)$$

$$+ \frac{1}{2}\frac{M^2}{A^2}\left(\frac{f_i q_i |q_i|}{\rho_i} + \frac{f_{i+1}q_{i+1}|q_{i+1}|}{\rho_{i+1}}\right) = 0 \tag{6.19}$$

and a similar expression for the energy equation.

Alternatively different equations can be evaluated at different grid points (Guy[16]) evaluating p at even-numbered grid points and q at odd-numbered grid points.

In any case a set of ordinary differential equations is derived

$$\mathbf{F}\left(\frac{d\mathbf{p}}{dt}, \mathbf{p, q, T}\right) = 0$$

$$\mathbf{G}\left(\frac{d\mathbf{q}}{dt}, \mathbf{p, q, T}\right) = 0 \qquad\qquad (6.20)$$

$$\mathbf{H}\left(\frac{d\mathbf{T}}{dt}, \mathbf{p, q, T}\right) = 0$$

where the ith elements of $\mathbf{F}$, $\mathbf{G}$, $\mathbf{H}$, will only depend on the ith and $(i + 1)$th elements of $\mathbf{p}$, $\mathbf{q}$, $\mathbf{T}$.

6.3.1.2. Finite element methods. The pipe is divided into several elements of length h, say. The pressure, flow and temperature are approximated by different polynomials in each element. There is a restriction that the approximate p, q, and T must be at least continuous. By choosing higher order polynomials one obtains a higher order of approximation. The pipeline equation (6.17) is expressed in the weak form

$$\int_{\text{pipe}} w\left[\frac{\partial p}{\partial t} - f(p, q, T)\right] dx = 0$$

$$\int_{\text{pipe}} w\left[\frac{\partial q}{\partial t} - g(p, q, T)\right] dx = 0 \qquad\qquad (6.21)$$

$$\int_{\text{pipe}} w\left[\frac{dT}{\partial t} - h(p, q, T)\right] dx = 0$$

These equations must hold true for all choices of weighting function w in order to make the bracketed quantities identically zero.

There is a choice of polynomial representation for p, q and T. The simplest representation is the linear or chapeau representation

$$\hat{p}(x) = \frac{x - ih}{h} p_{i+1} - \frac{(x - (i + 1)h)}{h} p_i$$

when $ih \leq x \leq (i + 1)h$.

It is not difficult to see that this can be written as

$$\hat{p}(x) = \sum_{i=0}^{n} p_i u_i(x)$$

where

$$u_i(x) = 0 \qquad\qquad x < (i-1)h$$

$$u_i(x) = \frac{x-(i-1)h}{h} \qquad (i-1)h \le x < ih \qquad\qquad (6.22)$$

$$u_i(x) = \frac{(i+1)h - x}{h} \qquad ih \le x < (i+1)h$$

$$u_i(x) = 0 \qquad\qquad x \ge (i+1)h$$

This has the shape of a witch's hat as in Fig. 6.3: hence the name chapeau.

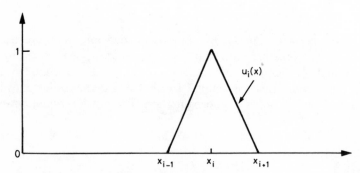

Fig. 6.3 *Chapeau function*

The set of $u_i(x)$ is called a basis. Plainly $\hat{p}(x)$ is an approximation to $p(x)$, and $\hat{p}(x)$ is continuous and differentiable everywhere, but the derivative is not continuous.

By grouping the grid points in three's instead of two's one can use a quadratic interpolation formula instead of Eqns. (6.22).

The Galerkin method consists of the following two steps:

(i) substitute $\hat{p}$ for p, $\hat{q}$ for q and $\hat{T}$ for T in the weak form of the equations, and

(ii) make a specific finite choice for $\{w\}$, the weighting functions in the weak form of the equations; namely, choose the basis functions themselves as weighting functions.

As shown in Rachford and Dupont[27] and Fincham and Goldwater[13] this procedure reduces the partial differential equations to ordinary differential equations. The procedure is rather lengthy and not widely used in gas network simulation.

The equations are non-linear and must of course be solved numerically.

Generally speaking using a higher order of basis function will give a more accurate solution but will involve more complicated ordinary differential equations to solve in time. However, when the solution possesses discontinuities high order methods may not always give more accurate solutions.

6.3.1.3 The method of characteristics. The previous methods merely involve translating the differential equations into algebraic analogues. The method of characteristics, however, involves changing coordinate systems.

Equations (6.14) and (6.15) can be linearly combined to give

$$
\lambda\left(\frac{\partial p}{\partial t} + \frac{1}{\lambda}\left(1 - \frac{M^2}{A^2}\frac{q^2}{\rho^2 a^2}\right)\frac{\partial p}{\partial x}\right) + \frac{M}{A}\left(\frac{\partial q}{\partial t} + \left(\lambda a^2 + \frac{2q}{\rho}\frac{M}{A}\right)\frac{\partial q}{\partial x}\right)
$$

$$
+ \frac{M^2}{A^2}\frac{2f\,|q|q}{\rho D} = 0 \quad (6.23)
$$

where λ is an arbitrary multiplier.

We will now choose λ so that the coefficients of $\partial p/\partial x$ and $\partial q/\partial x$ are equal

$$
\lambda = \left(-\frac{Mq}{\rho A a} \pm 1\right)\Big/ a = -\frac{v/a \pm 1}{a} = -\frac{m \pm 1}{a} \quad (6.24)
$$

The coefficients of $\partial p/\partial x$ and $\partial q/\partial x$ are then $(v \pm a)$, and Eqn. (6.23) becomes

$$
-\frac{m \pm 1}{a}\left|\frac{\partial p}{\partial t} + (v \pm a)\frac{\partial p}{\partial x}\right| + \frac{M}{A}\left|\frac{\partial q}{\partial t} + (v \pm a)\frac{\partial q}{\partial x}\right| + \frac{M^2}{A^2}\frac{2f\,|q|q}{\rho D} = 0
$$

$$
(6.25)
$$

The two signs give two equations which, of course, are exactly equivalent to Eqns. (6.14) and (6.15). We now consider two families of curves—the characteristics—C^+ and C^-. C^+ is given by

$$
\frac{dx^+}{dt} = v + a \quad (6.26)
$$

and C^- by

$$
\frac{dx^-}{dt} = v - a \quad (6.27)
$$

The equations above now become the characteristic equations

$$
-\left(\frac{m \pm 1}{a}\right)\left|\frac{\partial p}{\partial t} + \frac{dx^\pm}{dt}\frac{\partial p}{\partial x}\right| + \frac{M}{A}\left|\frac{\partial q}{\partial t} + \frac{dx^\pm}{dt}\frac{\partial q}{\partial x}\right| + \frac{M^2}{A^2}\frac{2f\,|q|q}{\rho D} = 0
$$

$$
(6.28)
$$

or

$$
-\left(\frac{m \pm 1}{a}\right)\frac{D^\pm p}{Dt} + \frac{M}{A}\frac{D^\pm q}{Dt} + \frac{M^2}{A^2}\frac{2f\,|q|q}{\rho D} = 0 \quad (6.29)
$$

where $D^\pm/Dt$ denotes the time derivatives following the characteristics. These equations are apparently a great simplification of Eqns. (6.14) and (6.15). Unfortunately the characteristics C^+ and C^- depend on the velocity of the gas and are thus, in general, not straight lines.

Numerical method of characteristics. The characteristic equations are simplified by ignoring the term $(M^2/A^2)\,\partial/\partial x(q^2/\rho)$. Assuming isothermal flow in a nearly ideal gas when $a^2 \doteq p/\rho = ZRT/M$ the characteristics become straight and Eqn. (6.25) reduces to

$$\frac{dx^\pm}{dt} = \pm a \tag{6.30}$$

$$\mp \frac{1}{a}\left(\frac{\partial p}{\partial t} \pm a\frac{\partial p}{\partial x}\right) + \frac{M}{A}\left(\frac{\partial q}{\partial t} \pm a\frac{\partial q}{\partial x}\right) + \frac{M^2}{A^2}\frac{2f\,|q|\,q}{\rho D} = 0 \tag{6.31}$$

The numerical method requires the use of Fig. 6.4.

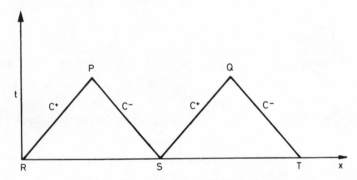

Fig. 6.4 *Diagram for method of characteristics*

Let the subscript R mean 'evaluated at $(0, 0)$; P at $(\Delta x, \Delta t)$; S at $(2\Delta x, 0)$; Q at $(3\Delta x, \Delta t)$; etc.' then Eqn. (6.29) can be integrated along a C^+ characteristic to give

$$-\frac{1}{a}\frac{p_P - p_R}{\Delta t} + \frac{M}{A}\frac{q_P - q_R}{\Delta t} + \frac{f}{D}\frac{M^2}{A^2}\left\{\frac{|q_P|q_P|}{\rho_P} + \frac{q_R|q_R|}{\rho_R}\right\} = 0 \tag{6.32}$$

and along a C^- characteristic to give

$$+\frac{1}{a}\frac{p_P - p_S}{\Delta t} + \frac{M}{A}\frac{q_P - q_S}{\Delta t} + \frac{f}{D}\frac{M^2}{A^2}\left\{\frac{|q_P|q_P|}{\rho_P} + \frac{q_S|q_S|}{\rho_S}\right\} = 0 \tag{6.33}$$

The last term is an approximation to

$$\frac{2}{\Delta t}\int_{C^\pm}\frac{M^2}{A^2}\frac{fq|q|}{\rho}\,dt$$

Equations (6.32) and (6.33) are a pair of simultaneous non-linear equations in just p_P and q_P and can fairly easily be solved in terms of p_R, p_S and q_R and q_S. Similarly all the other points like Q can be solved for p and q.

The characteristic lines C^+ and C^- are given by the equation $\Delta t = \Delta x/a$.

This illustration highlights the difficulties with curved characteristics. One would be solving for different levels of time at each step which is difficult in a computer program. Boundary conditions must be introduced to solve for the values at the boundary.

6.3.2 Actual models and computer programs—a review

Table 6.1 gives an idea of the approximations and steps taken by various people to produce a working model. One must make a distinction between the completeness of the physics one is modelling and the method used to solve the resulting equations.

In Table 6.1 a list of models and the main features of their approximations together with the methods of solution are compared. At the top are the most physically complete models with progressively more terms dropped towards the lower end. No account is given of models other than those which assume that the flow is one-dimensional and that viscous and turbulent effects can be calculated by using a Fanning friction factor.

Linebreak models. In order to estimate the potential hazard arising from a fractured pipe, estimates are sometimes needed of pipeline flow and pressures upsteam of a suddenly open pipe.

An analytic solution can be derived if the following assumptions are made:

(a) the energy equation is replaced by either the perfect gas law $p/\rho = RT/M$ or the adiabatic gas rule $p/\rho^\gamma = \text{const}$

(b) there is no friction.

This solution is the well-known shock-tube analysis—see, for example, Liepmann and Roshko.[21] It is worth studying the form of the solution which is shown in Fig. 6.5.

These graphs show that the gas expands as it comes out of the pipe. The expansion wave goes down the pipe with the velocity c and there is a discontinuity in the pressure gradient.

To derive a model for longer distances (say several tens of miles) a more complicated numerical model has to be used. Taylor[35] has used a model which incorporates the three conservation equations to give an adiabatic model in which the frictional reheating is accounted for. This uses an implicit finite difference approach with some allowance for the discontinuity in the derivatives. This model predicts that over a long distance the solution is mainly dependent on the frictional term rather than any inertia effects and the answers are, after a short time, comparable with a model which ignores inertia effects altogether.

Sens et al.[31] produced a similar model but flow was assumed to be isentropic. This assumption ignores frictional reheating of the gas and produces

Table 6.1 *Summary of models*

Model	Approximation	References
Linebreak	Adiabatic	Sens et al.,[31] Taylor[35]
Finite element	Isothermal	Rachford and Dupont[27]
Method of characteristics	Isothermal Neglect $\partial/\partial x(q^2/\rho)$	Stoner,[33] Streeter and Wylie,[34] Wylie et al.[42]
Finite differences/ Implicit—I	Isothermal Neglect $\partial/\partial x(q^2/\rho)$	Guy,[16] Heath and Blunt,[18] Sanchez,[29] Schmidt and Weimann,[30] Wylie et al.[41]
Finite differences/ Implicit—II	Isothermal Neglect $\partial/\partial x(q^2/\rho)$ and $\partial q/\partial t$	Goldwater et al.,[15] Larson and Wismer,[20] Michon et al.[22]
Finite differences/ Explicit	Isothermal Neglect $\partial/\partial x(q^2/\rho)$ and $\partial q/\partial t$	Goacher,[14] Distefano,[11] Tuppeck and Kirschke[37]
Steady state	Neglect all time derivatives	Buthod et al.[8] Hamam and Brameller,[17] Travers[36]

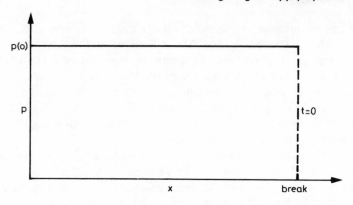

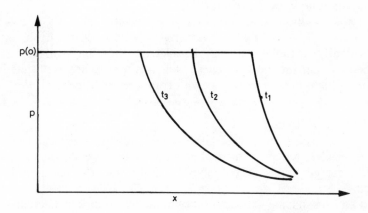

Fig. 6.5   *Pressure profiles for shock tube solutions*

temperatures which are too low, but the flows are not very different from Taylor's model.

Finite element models. Rachford and Dupont[27] used an isothermal finite-element model. When used to simulate ordinary gas transmission dynamics they claim high accuracy for the model. This is because they use cubic Hermite splines which should give errors which are $0(h^4)$ where h is the step-length.

However this accuracy cannot always be usefully realized, for example, on a network like the U.K. National Grid the geographical location of the offtakes constrains h to be generally small, and low-order methods are sufficiently accurate.

Methods of characteristic models. These models use the method of characteristics (MOC) to solve the equations of pipeflow. They neglect the part of the momentum term: $(M^2/A^2)\partial/\partial x(q^2/\rho)$ and thus get straight characteristics.

However, the method of characteristics is difficult to apply to a network because of the restriction $\Delta x = a \, \Delta t$. In Streeter and Wylie[34] the MOC is combined with an implicit finite-difference method and in Wylie et al.[42] the MOC is modified using a technique suggested by Yow.[43]

Finite difference models. Finite differences are by far the most popular method for modelling gas network dynamics. The majority of models neglect the term $(M^2/A^2)\partial/\partial x(q^2/\rho)$ as in the MOC, but some also neglect the term $(M/A)\partial q/\partial t$ to get a creeping flow model.

Steady-state models. If we set all time derivatives to zero we obtain a steady-state model for the network. These models consist of a set of algebraic equations as compared with the differential equations of dynamic models and can be solved in far less computer time. Steady-state models are used for low-pressure networks and for design work in high-pressure networks when linepack is not an important factor.

A popular method of solution for low-pressure networks is the Hardy Cross method (Travers[36]) which was originally designed for hand solutions. More recently a rapidly convergent Newton-Raphson method has been described by Hamam and Brameller[17] which is particularly well suited to the solution of high-pressure networks with compressors (Fincham[12]).

6.3.3 Time discretization

All methods, except for MOC, give rise to ordinary differential equations which themselves require numerical integration. Because of the large numbers of non-linear equations to be solved a simple method has usually to be adopted.

Most authors, having derived a set of non-linear ordinary differential equations from the spatial discretization, opt for one of the simple solution methods:

(a) forward difference (explicit method)
(b) central difference (Crank-Nicolson)
(c) backward difference (fully implicit method).

The solution of differential equations is well served in the literature (see, for example, Ames[1]) and we will not detail these methods here except for a few remarks. The first method given above, is very simple to program but suffers from stability problems; the second method is usually stable, has high-order accuracy but for sudden changes in forcing function (i.e. gas demands in gas network) the solution is prone to oscillate about the true solution; the third method gives absolute stability and no spurious oscillations but does not have the high asymptotic accuracy (as $\Delta t \to 0$) of the Crank-Nicolson method. The second and third methods involve the solution of simultaneous non-linear equations usually by Newton-Raphson linearization. Fortunately, for gas networks the Jacobians are sparse and full advantage of this must be taken to solve the large sets of equations economically. Most authors will have their own short cuts to solution of the equations, which contribute considerably to the economy of the process.

6.4 Networks

The main difficulty in extending the mathematical analysis to networks is the notation. We use here a notation which simplifies the algebra.

Mathematically a gas network can be thought of as having five classes of objects:

(a) Nodes which define the positions of the following four objects or which might be arbitrary points along a pipe.

(b) Pipes—these always lie between two distinct nodes and they will have length, roughness or efficiency, diameter and elevation. They will also have a direction which we say goes in a positive direction from the *left* node to the *right* node. This defines flow to be positive if the gas goes from the left node to the right node.

(c) Sources—these are the entry points for gas into the system. In practice they are often pressure controlled although more complex controls may be used. A source is always at a node.

(d) Demands—these are usually given as flow rates in standard volumes per unit time. Demands may in practice be sources if specified as a negative flow. Demands will always be at a node.

(e) Machines—these may be compressors or governors (pressure reduction stations) and their mode of control will depend on the network. They have a direction which is physically meaningful. For example, a normally functioning compressor will pump in a fixed direction and a governor will try to ensure that its outlet pressure is always lower than its inlet pressure. A pipe is merely a passive objective and a machine is an active objective.

In this chapter we will assume machines to lie between two nodes. Some authors assign them to one node and use various devices for assigning direction.

A network is thus an ordered set of nodes, pipes, sources, demands and machines. We use the following notation for sets of pipes incident upon the kth node:

S_k = set of pipes incident upon the kth node
L_k = subset of S_k for which the kth node is the left node
R_k = subset of S_k for which the kth node is the right node
$S_k = L_k \cup R_k$, of course.

Now let x_j be the distance along the jth pipe. We might want to divide this pipe into smaller lengths of h_j (say) in order to apply the finite difference formulations [Eqns. (6.18) and (6.19)] to each pipe.

In order to simplify the notation and algebra we will consider each such subdivision to be a single pipe, so that the points $x_j = nh_j(n = 0, \ldots, l_j/h_j)$ become nodes.

In the finite-difference formulations we need now only consider the pressures at the end of the pipe, the nodes, so that if pipe j lies between the kth node $(x_j = 0)$ and mth node $(x_j = l_j)$ the derivative along the pipe is approximated by

$$\frac{p(x_j = l_j) - p(x_j = 0)}{l_j}$$

or $(p_m - p_k)/l_j$.

If central values for the pipe are required, such as for $\partial p/\partial t$, then the approximation $\frac{1}{2}\{dp_m/dt + dp_k/dt\}$ can be used. These expressions can be substituted into Eqns. (6.18) and (6.19), changing the i into k, $i + 1$ into m, and h into l_j.

Continuity. Plainly unless there are shock waves (which we will not consider here) q and p will be continuous along a pipe of uniform diameter where there are no offtakes. When there is a sudden change of velocity, for instance just downstream of an offtake or at a multipipe junction, p is discontinuous; however, the dynamic head $p + \rho v^2$ is continuous. Nearly all authors neglect ρv^2 compared with p (see Section 6.3) and are therefore able to assign a unique nodal value to the pressure which will be p_k for the kth node.

There is also the mass balance equation at offtakes and multijunction nodes. If we take $q_j(x_j)$ to be the (continuous) molar flow along the jth pipe then for the kth node we have

$$- \sum_{j \in L_k} q_j(0) + \sum_{j \in R_k} q_j(l_j) = d_k \tag{6.34}$$

where d_k is the demand at the kth node.

Finally energy should be conserved at a node. This will certainly be the case if pressure and temperature are continuous at a node.

6.4.1 Active components of a network. Above we state that compressors, regulators, and valves can be active components of a network, that is they increase or reduce the flow and pressure of gas in a way different from a pipe. For a compressor, for example, there is a choice of equation. Often the machine will be controlled to keep the outlet pressure constant

$$p_O = p_{set}$$

or it might keep the compression ratio constant

$$(p_O/p_1) = r_{set}$$

In any case the outlet pressure, compression ratio, rotor speed, power and gas temperature must all be kept below prescribed or physical maxima. So a particular machine will be governed by a set of inequalities

(a) $p_O \leq p_{set}$ (pressure constraint)

(b) $p_O \leq rp_1$ (compression ratio constraint)

(c) $\{\kappa/(\kappa - 1)\}\beta(1 - (p_O/p_1)^{(\kappa - 1)/\kappa}) \leq W_{max}$ (power constraint)

(here κ = polytropic index 1.35 for natural gas and β depends on units and temperature of the gas).

The other equations become progressively more complicated and will not be listed here (see Cambell et al.[9]). Usually exactly one constraint will be active at any one time. A computer simulation program must regularly check these constraints.

In regulators there will be a different set of constraints which limit the outlet pressure, the gas velocity and pressure drop across the regulator. In valves there might be simply a one-way flow constraint. Wylie et al.[42] give a flow-pressure drop relationship for regulators and valves.

6.5 Application of simulation programs

There is a large range of problems in design and control of gas networks that can be solved with the aid of a gas-flow simulation program. The time scale covered by these problems extends from control problems a few hours ahead to the development of long-range strategy up to 20 years in the future. In short-term problems we should have available reasonably accurate data for the simulation program and must be able to predict ahead with commensurate accuracy. In the long-term problems we will not have accurate data and the accuracy of predictions will not be so important though we must be able to compare alternative strategies with some confidence.

6.5.1 Design applications

In design problems simulation is used to check that possible network designs satisfy the design criteria; for example, pressures should be greater than a specified minimum and compressors must be operating within design limits. There will probably be many grid designs that could meet specified demand levels and it is not usually practicable to check all these by using simulation, so the designer must select the most promising alternatives to compare. In practice further alternatives will probably occur to the designer during his study of the simulation results on his chosen alternatives. In this way it is possible to arrive at a good design which, though not necessarily the theoretical optimum, will be fairly close. By studying the behaviour of the system under conditions of failure of particular items of plant the designer can also ensure that his design has the required degree of robustness.

It is not always necessary to use dynamic simulation in design even in high-pressure networks. If the network is principally used for transmission of gas and not linepack storage then steady-state analysis can be used for a large part of the design work with a consequent saving in the amount of data and the computing time required. An example of a high-pressure design study using steady-state analysis can be found in Goldwater et al.[15] Low-pressure networks are traditionally designed on steady-state analysis because their storage capacity is very small.

In many high-pressure networks linepack is very important and is designed to be exploited in the operation of the network. For these networks dynamic simulation is essential. A number of papers discuss this application of simulation at some length: Ashworth and Towler,[2] Booth,[4] Cleeve et al.,[10] Hughes et al.,[19] Needham and Blunt,[23] Newbold[24] and Rachford.[25,26]

6.5.2 Control applications

A fair proportion of the application of simulation in control is concerned with studying problems weeks or months ahead. Typical applications in this category are in planning operations to set broad guidelines for day-to-day control or to cope with a failure of an item of plant, or calculating the effect of taking plant out for maintenance. These types of problems can be tackled in a similar way to design problems except that the number of alternatives to be considered is generally smaller—the network configuration is already fixed—and the main decision to be made is how it should be operated.

The short-term prediction of grid behaviour is a rather different proposition. In this application there is usually insufficient time for the controller to assemble the input data and to check through the results. There is a need for a high degree of automation in setting up the data and for a high degree of reliability and accuracy in performing the analysis. Automation is possible if the grid conditions are monitored by a control computer so that up-to-date data on grid behaviour will be available on that computer; this data is often referred to as 'on-line data' because it is being gathered as the computer monitors actual grid conditions. On-line pressure data is used to generate starting conditions for a prediction run; we also need data on flows to generate flow profiles, and these profiles must be combined with a demand prediction to produce the demand data required. The network configuration data must also be available and must be updated manually when changes are made, though some network data, such as whether compressors are on or off, can be determined from the on-line data. The only data that need be input manually at the time of the simulation are the information on the planned changes to be made to control variables; these may, of course, be changes that the controller is testing by use of simulation. This method of using simulation programs is referred to as on-line simulation because it uses on-line data. On-line simulation is used to predict up to 24 hours ahead. Though on-line simulation is being used its value is not yet fully established because there are some fundamental problems with the technique. One problem is getting accurate predictions and this is usually determined by the accuracy of the input data; all the data is subject to error and proper checks must be made to reduce this error. However, we cannot eliminate all error; for example, demand forecasts can, on occasions, be seriously wrong! Another problem is that to test alternative control strategies the controller may be required to perform several analyses, this can be time consuming, particularly as he must input data to define these control strategies, and may not be realistic when decisions must be made quickly under conditions of stress. Without using simulation the controller must be able to select a safe operating policy though this safe policy will probably be more expensive than some of the alternatives that could be found with simulation.

Ideally the control problem should be solved by optimization techniques without the need to resort to repeated simulations. Unfortunately because of the large number of variables and the non-linearity of the equations real gas networks are probably beyond the capacity of current techniques. However, a method for dynamic control of a network with optimization was presented by Larson and Wismer,[20] and Brameller et al.[6] presented a method for optimization of a transmission network for manufactured gas using a steady-state flow model.

Further discussions on dynamic simulation applied to control problems can be found in Bellars et al.[3] and Boyer.[5]

6.5.3 Model verification

Nearly all authors have reported tests on real networks to verify the accuracy of their models. The published results show that model pressure predictions are almost always within 5 % of measured pressures and generally much better. It should be remembered, however, that the models have 'adjustable' parameters (e.g. pipe efficiency) which can be chosen to give a better fit to the data than would be possible in normal simulation. Because proper tests are difficult and expensive to do they are generally limited to small sections of networks subject to low frequency disturbances. Such tests are not sufficiently sensitive to discriminate between the alternative physical models because simple creeping motion is a good approximation to these slow dynamics.

A test involving very rapid dynamic effects is reported by Sens et al.[31]; they vented gas in a high-pressure pipeline to atmosphere by opening a valve rapidly. They use this test to validate their linebreak model. A particular problem with measuring rapid dynamic effects is that the results are very sensitive to the time scale of the disturbing phenomena; this was demonstrated by Rachford and Dupont[28] in their discussion of simulating a compressor failure. Sens et al.[31] also noted that the pressure-time curves near the valve were strongly affected by the method of opening the valve, but these differences were not significant at a few kilometres distance from the valve.

6.6 References

1 AMES, W. F.: 'Numerical methods for partial differential equations', 2nd ed. (Nelson, 1977)
2 ASHWORTH, M., and TOWLER, J. H.: 'A simulation approach to transmission planning and control', *J. Inst. Gas Eng.*, 1972, **12**, 133–143
3 BELLARS, W. J., FRANCIS, R. F., and GUNNING, F.: 'Software for advisory control of the national transmission system', Gas Council Research Communication GC 197, 1972
4 BOOTH, P. G.: 'Diurnal storage planning and the use of transient analysis', *J. Inst. Gas Eng.*, 1975, **15**, 380–385
5 BOYER, H. M.: 'Transient flow simulation as a training aid for gas dispatchers', Paper presented at the American Gas Association transmission conference, St. Louis, May 16–18, 1977
6 BRAMELLER, A., CHANCELLOR, V. E., HAMAM, Y., and YALCINDAG, C.: 'Recent advances in mathematical analysis for gas industry application', *J. Inst. Gas Eng.*, 1971, **11**, 188–219
7 BROWN, F. T., MARGOLIS, D. L., and SHAH, R. P.: 'Small amplitude frequency behaviour or fluid lines with turbulent flow', *J. Basic Eng.*, Trans. ASME Series D, 1969, **91**, 678–692
8 BUTHOD, A. P., CASTILLO, G., and THOMPSON, R. E.: 'How to use computers to calculate heat, pressure in buried pipelines', *Oil and Gas J.*, 1971, **69**, 57–59
9 CAMBELL, R. V., VANDAVEER, F. E., and VEIT, C. J.: 'Compressor stations' in 'Gas Engineers Handbook' (Industrial Press, 8/50–8/91, 1965)
10 CLEEVE, P., HARTILL, I. J., and WADE, V. A. H.: 'Southerngas Engineering Planning', *J. Inst. Gas Eng.*, 1977, **17**, 143–155
11 DISTEFANO, G. P.: 'PIPETRAN, version IV, A digital computer program for the simulation of gas pipeline network dynamics', Cat. No. L20000 American Gas Association Inc., New York, 1970
12 FINCHAM, A. E.: 'A review of computer programs for network analysis', Institute of Gas Engineers communication GC 189, 1971
13 FINCHAM, A. E., and GOLDWATER, M. H.: 'Simulation models for gas transmission networks', *Trans. Inst. Measurement and Control*, 1979, **1**, 3–12
14 GOACHER, P. S.: 'Steady and transient analysis of gas flows in networks', *J. Inst. Gas Eng.*, 1970, **10**, 242–264
15 GOLDWATER, M. H., ROGERS, K., and TURNBULL, D. K.: 'The PAN network analysis program—its development and use', Institute of Gas Engineers communication 1009, 1976
16 GUY, J. J.: 'Computation of unsteady gas flow in pipe networks' Proceedings of symposium, 'Efficient methods for practising chemical engineers', Symposium series No. 23, London Institute of Chemical Engineers, 139–145, 1967
17 HAMAM, Y. M., and BRAMELLER, A.: 'Hybrid method for the solution of piping networks', *Inst. Elect. Eng. Proc.*, 1971, **118**, 1607–1612
18 HEATH, M. J., and BLUNT, J. C.: 'Dynamic simulation applied to the design and control of a pipeline network', *J. Inst. Gas Eng.*, 1969, **9**, 261–279
19 HUGHES, H. W. D., LANGFORD, R. G., and ROUSE, M. J.: 'Practical application of transient flow analysis', Institute Gas Engineers communication 881, 1972
20 LARSON, R. E., and WISMER, D. A.: 'Hierarchial control of transient flow in natural gas pipeline networks', Proc. of IFAC Symposium 'Control of distributed parameter systems', Banff, Canada Paper 6-1, 1971
21 LIEPMANN, H. W., and ROSHKO, A.: 'Elements of Gasdynamics' (John Wiley, New York, London and Sydney, 1957)
22 MICHON, R., SORINE, M., and SOULAS, M.: 'Modèle numérique de calcul de l'ecoulement des gaz en régime variable', Association Technique de l'Industrie du Gaz en France Congrès, 1978

23 NEEDHAM, D., and BLUNT, J. C.: 'Pipeline networks design for transient flow conditions', Paper IGU/C 15-70, 11th International Gas Conference, Moscow, 1970

24 NEWBOLD, J.: 'The role of storage mains in the operation of the gas supply system', *J. Inst. Gas Eng.*, 1974, **14**, 271-280

25 RACHFORD, H. H., Jr.: 'Transient pipeline calculations improve design', *Oil and Gas J.*, 1972, **70**(44), 54-57

26 RACHFORD, H. H., Jr.: 'Easy-to-use transient models can answer critical pipeline questions', *Oil and Gas J.*, 1972, **70**(45), 64-66

27 RACHFORD, H. H., Jr., and DUPONT, T.: 'A fast highly accurate means of modeling transient flow in gas pipeline systems by variational methods', *Soc. Pet. Eng. J.*, 1974, **14**, 165-178

28 RACHFORD, H. H., Jr., and DUPONT, T.: 'Some applications of transient flow simulation to promote understanding the performance of gas pipeline systems', *Soc. Pet. Eng. J.*, 1974, **14**, 179-186

29 SANCHEZ, F.: 'Analysis de una red de distribution de gas en regimen variable', *Rev. Inst. Mex. Pet.*, 1977, **9**, 20-25

30 SCHMIDT, G., and WEIMANN, A.: 'Instationäre Gasnetzberechnung mit dem Programmsystem GANESI', *GWF-Gas/Erdgas*, 1977, **118**, 53-57

31 SENS, M., JOUVE, Ph., and PELLETIER, R.: 'Détection d'une rupture accidentelle de conduite' Paper IGU/C 37-30 presented at the 11th International Gas Conference, Moscow, 1970 (English translation: British Gas Internal Report LRS T448)

32 SHAPIRO, A. H.: 'The dynamics and thermodynamics of compressible fluid flow' (Ronald Press Co., New York, 1954)

33 STONER, M. A.: 'Analysis and control of unsteady flows in natural gas piping systems', Trans ASME, Series D, *J. Basic Eng.*, 1960, **91**, 331-340

34 STREETER, V. L., and WYLIE, E. B.: 'Natural gas pipeline transients', *Soc. Pet. Eng. J.*, 1970, **10**, 357-364

35 TAYLOR, B. A.: 'The flow in pipelines following catastrophic failure', British Gas Internal Report LRS 338, 1978

36 TRAVERS, K.: 'The mesh method in gas network analysis', *Gas J.*, 1967, **332**, 167-174 (1st Nov.)

37 TUPPECK, F., and KIRSCHKE, H.: 'Ein numerisches Verfahren zur Berechnung instationärer Strömungsvorgänge in Ferngasleitungen', *GWF-Gas/Erdgas*, 1962, **103**, 523-528

38 UHL, et al.: 'Steady flow in gas pipelines', American Gas Association, Inc. New York, 1965

39 WARD SMITH, A. J.: 'Pressure losses in ducted flows' (Butterworths, London, 1971)

40 WEIMANN, A.: 'Modellierung und Simulation der Dynamik von Gasverteilnetzen im Hinblick auf Gasnetzführung und Gasnetzübertwachung', Dr. Ing. Thesis, Munich Technical University, 1978 (English Translation: British Gas Internal Report LRS T435)

41 WYLIE, E. B., STREETER, V. L., and STONER, M. A.: 'Network system transient calculations by implicit method', *Soc. Pet. Eng. J.*, 1971, **11**, 356-362

42 WYLIE, E. B., STREETER, V. L., and STONER, M. A.: 'Unsteady-state natural gas calculations in complex pipe systems', *Soc. Pet. Eng. J.*, 1974, **14**, 35-43

43 YOW, W.: 'Numerical error in natural gas transient calculations', Trans. ASME, Series D, *J. Basic Eng.*, 1972, **94**, 422-428

Chapter 7

Modelling of coal and mineral extraction processes

J. B Edwards

7.1 Introduction

Mining involves the following fundamental operations:

(a) The actual winning of the coal, stone or mineral ore from the underground seam or vein in which it naturally occurs.

(b) The clearance of the material from several such sources in any one mine to some central point for the final operation.

(c) The preparation of the extracted raw material for sale, involving perhaps washing, crushing and grading or blending according to lump size or composition.

This chapter will concentrate on those aspects of operations (a) and (b) which involve process modelling for the design of controls, the satisfactory performance of which is essential to efficient mining. Coal and mineral preparation processes, (c), tend to be manageable by conventional process control techniques and will therefore not be studied in detail here. It is within the areas of winning and clearance that process models have characteristics which are, at first sight, peculiar to mining and therefore deserving of a special chapter devoted to their development. Important points of contact with process models outside the mining field will also become apparent, however, thereby hopefully contributing to the art of process modelling generally.

Of course an enormous number of activities are also involved in mining without which the three basic operations (a), (b), or (c) could not take place. These include, for instance, mineral location by exploration and drilling, marketing, mine development, plant maintenance, the supply of services such as electricity, compressed air, men, materials and ventilation to all points in the production network, the removal of water and also unsaleable rock (necessarily extracted to provide underground roadways of a size adequate for the free flow of product and services throughout the mine). Indeed every underground process from the hoisting of material from the pit bottom to the supply of power to the many hundreds of machines could fairly be termed a 'mining process' in its own right despite the existence of apparently related processes in industry generally. What makes the mining versions special are:

(i) The enormous distances and depths over which the processes operate.

(ii) The massive nature of the machines involved to withstand the arduous underground environment which imposes the most severe stresses from heat, rockfall, water and dust and, in coal mining, the need for flameproofness of electrical equipment to avoid methane ignition.

(iii) The non-static nature of mining which must constantly chase a retreating product so necessitating the continual extension and evolution of the back-up systems required for its extraction.

Space does not, unfortunately, permit investigation of these subsidiary but essential processes.

In mining, as in all productive industries, the key dependent variables requiring control are product quality and the rate of production. The attainment of high levels for both simultaneously often imposes a compromise dictated by market factors, costs and safety considerations. In this chapter we shall therefore devote attention to aspects of the winning and clearance operations which crucially affect the quality and quantity of the mineral extracted. The dynamics of cutting will be investigated since this process affects the lump size, and therefore the quality, of the product whilst having an important bearing on the rate of which cutting machines can progress. The steering of these machines through the strata will also receive attention since it is by successful steering that the volume extracted of the desired product can be maximized without risk of contamination resulting from the accidental cutting of unwanted rock strata above and beneath (or to either side of) the seam (or vein) of the desired mineral. If the seam, or vein, is softer than the surrounding strata then clearly successful steering again assists the increase of bulk production rate.

Some adjustments to product quality can, of course, be made after the initial winning of the product by crushing, separation, and blending etc., either in the product-preparation phase of operations or, at intervals, in the clearance system. The prime purpose of the clearance system is, however, the rapid removal of the product through the mine to the surface without the development of spillage or bottlenecks, and the modelling of such systems to achieve this goal by proper control will be examined.

There exists throughout mining a great variety in the types of individual machines and machine systems employed and a comprehensive coverage cannot be accomplished within the confines of a single chapter. The description of only some of the types in merely qualitative terms could easily occupy an entire volume. The aspects considered and the approach to modelling the particular examples (drawn largely from coal mining), here presented will nevertheless be found to apply in most mining situations. Furthermore, the behavioural characteristics predicted for these specific model examples will also be found to apply, at least qualitatively, to most related mining systems which will differ only in detail, and not in basic principle of operation, from the examples considered.

7.2 The dynamics of coal and mineral cutting

Cutting machines, in principle at least, resemble machine tools designed for the shaping of metals and, despite some important differences such as substantial frictional effects, a good deal can be learned from attempts[1, 2] to model their cutting actions that is useful in predicting aspects of the behaviour of coal—rock—and mineral-cutting machines. Like machine tools they incorporate two fundamental mechanisms which are: (i) a cutting head, usually rotary in operation, the cutting picks of which tear material from the rock- or coal-face; and (ii) a feed- or, in mining terminology, haulage-mechanism provided to advance the cutting head linearly into the solid rock or coal as the previously fractured material is continuously cleared.

To prevent the uncontrolled build-up of cut material around the cutting head, this unit is usually equipped with vanes or paddles which fling the cuttings onto some form of automatically advancing conveyor which, whilst being intimately linked to the cutting machine, may be regarded as the first link in the conveying chain which constitutes the clearance system.

Figure 7.1 illustrates the cutting action of a shearer or milling machine in which the cutting drum or disc rotates about an axis perpendicular to the

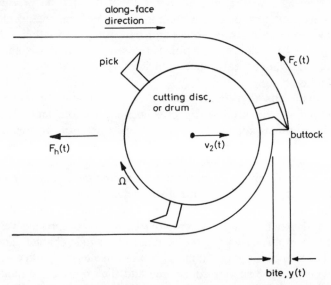

Fig. 7.1   *Showing cutting action of a miller or shearer machine*

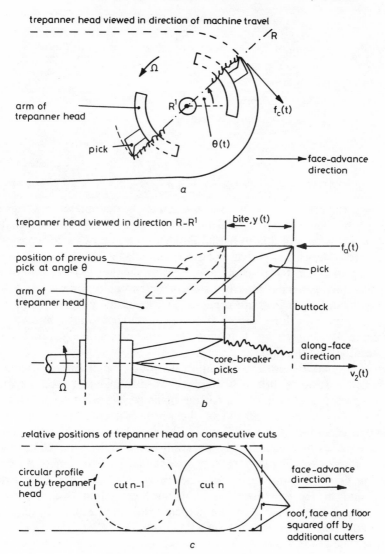

trepanner head viewed in direction of machine travel

arm of
trepanner head

pick

Ω

R^1

$\theta(t)$

R

$f_c(t)$

face-advance
direction

a

trepanner head viewed in direction R-R¹

bite, y (t)

position of previous
pick at angle θ

arm of
trepanner head

Ω

pick

buttock

core-breaker
picks

along-face
direction

$f_a(t)$

$v_2(t)$

b

relative positions of trepanner head on consecutive cuts

circular profile
cut by trepanner
head

cut n-1

cut n

face-advance
direction

roof, face and floor
squared off by
additional cutters

c

Fig. 7.2 *Trepanner cutting action*

direction of feed or haulage whereas a trepanner head rotates about an axis parallel to the haulage direction as shown in Fig. 7.2. A diagrammatic plan view of a shearer machine is shown in Fig. 7.3 which illustrates a machine with an inbuilt haulage system which pulls the machine along a fixed rope or chain as shown. As an alternative, an external winch may be used for haulage purposes situated at one or other end of the face. In this case the cutting machine is hauled by a moving loop of chain (or rope) which passes round the powered haulage drive sprocket at one end of the face and an idler

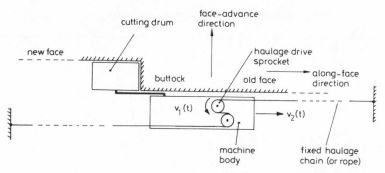

Fig. 7.3   *Plan view of shearer machine showing inbuilt haulage system*

sprocket at the other. The stiffness of the haulage system therefore varies with the length of chain or rope in tension—which varies from one face length to near-zero in the case of inbuilt haulage and from either two to one (or one to near-zero) face lengths in the case of external haulage—depending on the side of the loop to which the machine is attached. In either case a chain or rope type of haulage system is far from stiff for much of the time because of the long lengths of face† which are generally employed. Because of the potential hazard presented by long, highly stretched ropes and chains, and to avoid control problems posed by their varying stiffness, rack-and-pinion haulages have recently been developed.

Figures 7.2 and 7.3 also serve to illustrate how, in so called *longwall mining* the mineral is removed not by a simple forward tunnelling procedure but by repeated passes or sweeps of the cutting machine, exposing a new face with each pass. The machine, and indeed the whole face installation, comprising roof supports, conveyor and services, is advanced by hydraulic jacks between successive passes by a distance W_d, generally equal to the width of the cutting drum or disc or the diameter of, say, a trepanner head to ensure maximum extraction of mineral on each pass. The direction of machine travel, whilst cutting, and the general direction of mineral extraction over a period of several passes, or cuts, are therefore perpendicular to one another, the former being known as the *along-face* direction and the latter, the *face-advance* direction. These directions are indicated in Figs. 7.1, 7.2 and 7.3. Provided the cutting machine with its supporting installation is fully advanced between cuts, then the multipass nature of the longwall process has little direct bearing on the cutting process as such but has a profound effect on machine steering as will be shown in Section 7.3.

7.2.1 Modelling the cutting head
The rotary cutting head is fitted, around its periphery, with a number of cutting picks, which are responsible for tearing the coal or mineral from the

---

† A fuller discussion of so called '*longwall*' mining will be found in the introduction to Section 7.3.

solid face and which therefore absorb the bulk of the power supplied to the cutting machine through its haulage and cutting-head drive motors. The circumferential forces on these picks are provided by the cutting-head drive whilst the net horizontal force† on the head, arising from the vectorial summation of all the radial pick-force components in the case of a shearer or miller (or the total of the axial force components in the case of a trepanner) is provided by the haulage. Any resultant of the sideways forces on the cutting picks will produce friction forces between the machine and either the face-wall or the floor or track upon which the machine slides, which in turn will be overcome by a small proportion of the power available from the head and haulage drive motors.

Now in fundamental laboratory experiments[3, 4, 5] on individual cutting picks it has been shown that all three components of the cutting force depend predominantly on the following factors:

(a) the geometry, general design and sharpness of the pick itself
(b) the *bite*, or *penetration* of the pick into the material
(c) the hardness, H, of the material being cut.

Furthermore, for a given type and sharpness of pick, the force/bite relationship is substantially linear over a wide range of penetrations. For the entire cutting head therefore it may be assumed that the net horizontal (along-face) and circumferential forces $F_h(t)$ and $F_c(t)$ are given by

$$F_h(t) = k_h(H)y(t) \tag{7.1}$$

and

$$F_c(t) = k_c(H)y(t) \tag{7.2}$$

where k_h and k_c are machine and mineral constants varying, for a given machine, with the mineral hardness and pick sharpness. The variable $y(t)$ is the *effective bite* taken by the cutting head.

Now consideration of Fig. 7.2 would indicate that, in the case of a trepanner head, $y(t)$ is governed by the differential-delay equation

$$dy(t)/dt = v_2(t) - v_2(t - T_d) \tag{7.3}$$

where v_2 is the speed of translation of the cutting machine along the face and T_d is the interval between the arrival of consecutive picks, being given by

$$T_d = \theta_p/\Omega \tag{7.4}$$

where θ_p is the angular spacing of the picks and Ω the angular speed of the head. T_d is virtually constant, cutting heads being driven generally by squirrel-cage induction motors. With a miller or shearer type of cutter, as illustrated in

† More precisely this force should perhaps be described as the net along-face cutting force to allow for inclined seams.

Fig. 7.1, $y(t)$ is also given by Eqn. (7.3) but the disc or drum is shown, in the figure, not at a general point but at a particular point in its rotation, viz. at the point of maximum bite: even with a constant speed v_2, the instantaneous bite will clearly vary as the drum rotates, being zero at top and bottom. To merely substitute Eqn. (7.3) into (7.1) and (7.2) to obtain the instantaneous cutting forces is therefore clearly an approximation in the case of shearer and milling machines.

That *bite* and *hardness* should be the principal variables affecting cutting forces is clearly reasonable in view of the experimental results for single picks but the analytical generalization of these findings to the multipick, field situation is not completely straightforward, partly for the reasons given above and partly because the leading pick of a group may relieve considerably the load imposed on those immediately following by breaking off a lump larger than its penetration prior to mineral fracture. Groups or lines of picks are therefore often best regarded as single *effective picks*. A shearer drum, for instance, as illustrated in Fig. 7.4, is usually laced with, i,—typically two or three spiral

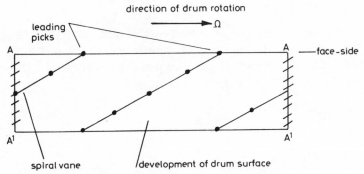

Fig. 7.4 *Spiral pick-lacing pattern for a shearer drum*

lines of picks to assist the clearance of the cut material, and such a drum is often best regarded conceptually not as a number of miller discs assembled side by side with staggered picks, but as a single miller disc having only i effective picks since the cutting load tends to be concentrated on the face-side of the drum.

Equations (7.1) to (7.3) therefore provide a reasonable conceptual basis for dynamic modelling but care is needed in establishing the number of effective picks, and therefore the effective value of T_d, and in predicting values for k_h and k_c from single-pick data. Confirmatory field experiments clearly play an important role in any novel situation. In the case of the trepanner, least error is involved in the use of Eqn. (7.3). However, if the object of modelling is to determine, say, the total torque or power consumption of the cutting head or haulage drive, then the necessity for these motors to drive additional jibs and discs to square off the original cylindrical excavation cut by a trepanner head introduces complications. The effect of the breaker picks (see Fig. 7.2) for shattering the core cut by the head would also require investigation for really rigorous model development. Since these operations tend to consume a relatively small proportion of the available power, however, they may be reasonably disregarded in the structure of the conceptual model and accounted for by appropriate increases in the values of parameters k_h and k_c. The values of k_h and k_c are comparable for conventional pick and head designs which means that the ratio of the power ratings for head and haulage drive is dictated largely by the ratio $\Omega.r./\bar{v}_2$, where r is the radius of the cutting head and $\bar{v}_2$ the mean value of $v_2(t)$. Machines such as shearers and trepanners designed to extract deep webs (i.e. large W_d) are generally slow moving along the face and hence their cutting-head drives draw considerably more power than do their haulages, typical machine ratings being 200 and 20 kW respectively. The haulages of fast-traversing millers on the other hand may absorb a considerably larger fraction of the total power used.

7.2.2 Modelling the haulage/transmission system

Because of the resilience of the transmission, the instantaneous translational speed $v_1(t)$ of the haulage/transmission drive may differ considerably from the machine speed $v_2(t)$, particularly in the case of rope- or chain-hauled machines. If, therefore, $\tau(t)$ denotes the tension in the transmission then

$$d\tau(t)/dt = k_a(l)\{v_1(t) - v_2(t)\} \tag{7.5}$$

where k_a is the stretch modulus of the transmission which, as indicated in Eqn. (7.5) will be dependent upon the length, l, of rope or chain in tension, in such a system. Generally the average speed of haulage is sufficiently small for any change in l to be negligible within the settling time or oscillation period of the slowest system mode so permitting dynamic simulations to be carried out at constant values of k_a pertaining to various points along the face. In the case of stiff transmissions exhibiting negligible deviation between v_2 and the speed of the final drive member, some resilience between v_2 and the *set speed* of the drive will generally occur nevertheless, due usually to the compressibility of hydraulic fluid. Interpreting v_1 as the set speed in such situations therefore permits the use of Eqn. (7.5), k_a now being position-invariant and related to the stiffness of the haulage drive itself.

The tractive effort $\tau(t)$ applied by the haulage/transmission to the cutting machine must provide not only the total resolved along-face cutting force $F_h(t)$ but must also overcome static and sliding frictional forces between the machine body or skids and the conveyor track or rough floor upon which the machine slides. Any linear acceleration forces required must also be provided by $\tau(t)$. We therefore have the force balance equation

$$\tau(t) = F_h(t) + F_f(t) + M \, dv_2/dt \tag{7.6}$$

where $F_f(t)$ is the frictional force given by

$$F_f(t) = \{\text{signum } v_2(t)\}F_s \qquad |v_2(t)| > 0 \tag{7.7}$$

and

$$F_f(t) = \tau(t) - F_h(t) \qquad v_2 = 0 \quad \text{and} \quad |\tau(t) - F_h(t)| < F_b \tag{7.8}$$

where F_s and F_b are the values of sliding and static frictional forces respectively, F_b being greater than F_s and both positive. The parameter M is the mass of the machine lumped together with one-third that of the rope or chain, the frequencies of disturbances transmitted along the rope or chain not usually being high enough to warrant the use of a distributed mass/elasticity model.

7.2.3 Modelling the cutting-head drive

Except in the case of high-speed millers where haulage power may exceed disc power, it is generally necessary to regulate the power or current consumption of the cutting-head drive (the power rating of which lies generally in the range 100 to 400 kW), by manipulation of the adjustable haulage drive speed $v_1(t)$ as the hardness H of the coal, rock or mineral varies. This drive is usually a squirrel cage induction motor of substantially constant speed as has already been mentioned but exhibiting some speed droop with increase in load nevertheless. Whilst insufficient to seriously affect the value of T_d (Section 7.2.1), in conjunction with the inertia of the drive motor, the speed droop does give rise to a significant time constant T_m in the response $P_e(t)$, the electrical power drawn by the motor, to disturbance $P_m(t)$, the mechanical power delivered to the cutting head. Within the normal working speed range of the motor we therefore have that

$$T_m \, dP_e(t)/dt = k_e P_m(t) - P_e(t) \tag{7.9}$$

where k_e (> 1) is a parameter related to the drive efficiency which may be assumed constant over the normal load range. The mechanical power $P_m(t)$ may in turn be related to the net circumferential cutting force $F_c(t)$ thus

$$P_m(t) = k_d F_c(t) \tag{7.10}$$

where k_d is a machine parameter which, whilst depending on rotational speed Ω, may again be taken as constant in studies of perturbations around normal load conditions.

7.2.4 The overall system model behaviour

By taking Laplace transforms, in s with respect to t, of the foregoing differential equations and assembling all the resulting algebraic equations, a block diagram of the composite system may be readily derived. Such a diagram is shown in Fig. 7.5 which also includes the representation of a closed-loop controller for the regulation of $P_e(t)$ to a reference value P_{ref}. As indicated,

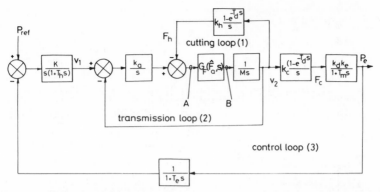

Fig. 7.5   *Block diagram of automatically controlled cutting and haulage system*

measurement of P_e usually involves the smoothing action of a low-pass filter, of time-constant T_e, to remove the alternating component of P_e and an actuator lag T_h is also involved in the manipulation of the set speed $v_1(t)$. The controller shown has a gain K and is of the pure integral type in accordance with usual practice. In Fig. 7.5, the non-linear relationship between machine speed v_2 and net applied force $\tau - F_h$ is represented by the describing function G_F/Ms which will be discussed later in connection with analytical predictions of system behaviour. For the moment we shall devote attention to computed solutions obtained directly from Eqns. (7.1) to (7.10).

The system's open-loop response to a step change in $v_1(t)$, at constant H, first reported[6] by the author in 1978, along with much of the material of Section 7.2 is shown in Fig. 7.6 for the following realistic system parameters

$$v_1(t) = 0.0762 \text{ m/s, increasing to } 0.1270 \text{ m/s}$$

$$k_a = 12700 \text{ kg/m } (\equiv l = 300 \text{ m})$$

$$k_c = k_h = 68800 \text{ kg/m}$$

$$T_d = 0.4 \text{ s}$$

$$F_s = 2270 \text{ kg}$$

$$F_b = 4540 \text{ kg}$$

$$M = 5120 \text{ kg}$$

From Fig. 7.6 it is clear that, in the presence of the cutting force F_h, the behaviour of v_2 is very different from that which would be expected if v_2 were restrained by friction and inertia alone. Under such circumstances it would be expected that for changes in $v_1 \ll v_{2\,max}$, the size and shape of the v_2-pulse should be little affected by the change in v_1 and the change should result instead in a change in the frequency of hopping. Indeed the value of $v_{2\,max}$ in the absence of F_h would be approximately $(F_b - F_s)/(k_a M)^{0.5} = 0.88$ m/s for

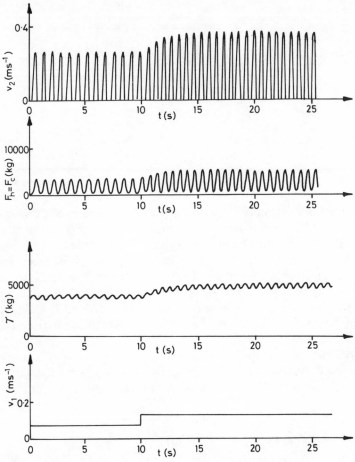

Fig. 7.6 *Simulated performance of shearer machine (open-loop control)*

the given parameters, compared to the much smaller values of Fig. 7.6. The increase in v_1 in fact causes only a marginal increase in frequency but produces a substantial change in $v_{2\,max}$ as shown. The frequency f of hopping always lies in the range

$$0 \cdot 5 T_d^{-1} < f < T_d^{-1} \tag{7.11}$$

and is therefore dictated by drum rotation rather than mere stick/slip phenomena which would produce a wide variation in the frequency of hopping with changing v_1. This hopping at near effective-pick frequency is a well-known characteristic of real machines and therefore its prediction provides a good measure of confidence in the model.

It is important also to notice that the tension τ varies little in the presence of F_h whereas it would undergo wild fluctuations in the presence of friction

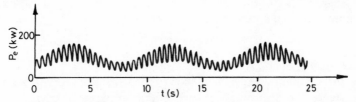

Fig. 7.7   *Simulated power consumption of cutting-drum drive (automatically controlled)*

alone. By contrast, the hopping, or chatter—to use the analogous term in machine-tool dynamics—exhibited by v_2 is strongly manifested in the F_c-, and therefore the P_m-trace, with important practical implications since stall values are easily attained if the mean value of P_e is regulated close to the continuously rated value† for the motor.

Load oscillations are in fact made worse by the addition of a responsive integral controller as predicted by Fig. 7.7 which is obtained from the simulation of the identical machine as for Fig. 7.6, with cutting-head drive and control-system parameters as follows:

$$v_{1\,max} = 0 \cdot 203 \text{ m/s}$$

$$K = 0 \cdot 0051 \text{ m/s/kW}$$

$$k_d = 0 \cdot 0355 \text{ kW/kg}$$

$$k_e = 1 \cdot 25$$

$$T_m = 0 \cdot 05 \text{ s}$$

$$T_e = T_h = 0 \cdot 10 \text{ s}$$

Low-frequency limit-cycling of the control system is now superimposed on all traces to varying degrees and clearly accentuates the problem of stalling. By way of confirmation, Fig. 7.8 shows an actual field trial recording of P_e

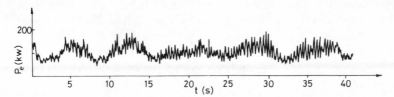

Fig. 7.8   *Field recording of shearer drum drive power consumption (automatically controlled)*

obtained on the same shearer machine as that for which the foregoing parameters were estimated. The high- and low-frequency components are clearly comparable in amplitude and frequency with those predicted and it was found necessary in practice to reduce P_{ref} to about 75% of the rated value of P_e to

† Typically the stall value of $P_e = 2$ to $2 \cdot 5$ times the continuous rating of an induction motor.

avoid regular stalling of the motor and the consequent loss of production and risk of motor burnout through restarting too frequently.

7.2.5 Analytical predictions

Whereas the simulation experiments for particular cases seem to accord with field experience it is also important to be able to make at least approximate predictions of system behaviour from the theoretical model without recourse to computer simulation, if only to restrict the number of simulation experiments otherwise necessary to obtain a comprehensive picture of system performance under a wide range of load conditions and parameter values. It is in the frequency domain that such analytical solutions are generally most readily obtained (see Volume I, Chapter 2), so requiring a model formulated in terms of interconnected transfer-functions as Fig. 7.5. Unfortunately the frictional effects described by Eqns. (7.7) and (7.8) are highly nonlinear so that the operator $G_F(s)$ shown in Fig. 7.5 relating the net applied force $F_a\,(= \tau - F_h)$ to the acceleration force $M\,dv_2/dt$ will be highly dependent upon its input signal amplitude $\hat{F}_a$ so that G_F is properly denoted as $G_F(\hat{F}_a, s)$. Of course $G_F(\hat{F}_a, s)$ strictly describes only the effect of the frictional process on the fundamental component of an oscillating input signal of non-sinusoidal wave shape and the omission of parallel transfer-functions processing any higher frequency harmonics is justified only if it can be assumed that these are effectively attenuated in passing round the feedback loops containing the nonlinear operation. The integration process $1/Ms$ immediately following the so-called describing function $G_F(\hat{F}_a, s)$ in Fig. 7.5 should certainly aid the attenuation of harmonics in this case. Before proceeding therefore it is necessary to determine the nature of G_F by considering a sinusoidal force $\hat{F}_a \sin \omega t$ applied to the friction-restrained machine.

7.2.5.1 An approximate describing-function representation of friction. Figure 7.9 illustrates the response $v_2(t)$ of a machine subject to a sinusoidal force $F_a(t) = \hat{F}_a \sin \omega t$ and restrained by static and sliding friction described by Eqns. (7.7) and (7.8). As indicated the motion will be discontinuous if $\hat{F}_a$ is small enough to ensure that the angle θ_b of breakaway and the angle θ_s at which the machine again becomes stationary are separated by less than half a half-cycle of oscillation, i.e.

$$\theta_s < \pi + \theta_b \tag{7.12}$$

For any motion to take place of course we must have that

$$\hat{F}_a > F_b \tag{7.13}$$

Now θ_b is given by

$$\theta_b = \sin^{-1}(F_b/\hat{F}_a) \qquad 0 < \theta_b < \pi/2 \tag{7.14}$$

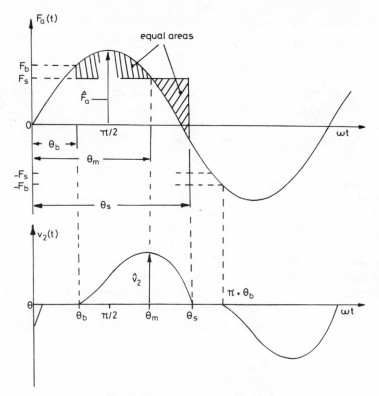

Fig. 7.9   *Discontinuous oscillation*

and θ_s can be obtained by integrating the equation of motion, viz.

$$M \, dv_2/dt = \hat{F}_a \sin \omega t - F_s \qquad \theta_b < \omega t < \theta_s \qquad (7.15)$$

and setting $v_2(t)$ to zero so that

$$(F_s/\hat{F}_a)(\theta_s - \theta_b) = \cos \theta_b - \cos \theta_s \qquad (7.16)$$

Now maximum velocity $\hat{v}_2$ will occur when $\omega t = \theta_m$ where

$$\theta_m = \pi - \sin^{-1} (F_s/\hat{F}_a) \qquad (7.17)$$

and, since integration of the equation of motion, (7.15), gives

$$v_2(t)\omega M = \hat{F}_a(\cos \theta_b - \cos \theta) + F_s(\theta_b - \theta) \qquad (7.18)$$

where

$$\theta = \omega t \qquad (7.19)$$

MDS - O

on substituting θ_m for θ we get, using Eqns. (7.14) and (7.17),

$$\hat{v}_2 \omega M/\hat{F}_a = \{1 - (F_s/\hat{F}_a)^2\}^{0.5} + \{1 - (1 - F_b/\hat{F}_a)^2\}^{0.5}$$
$$+ (F_s/\hat{F}_a)\{\sin^{-1} (F_s/\hat{F}_a) + \sin^{-1} (F_b/\hat{F}_a) - \pi\} \qquad (7.20)$$

so yielding an expression for the velocity amplitude in terms of the applied force parameters $\hat{F}_a$, ω and system parameters. Turning attention to the relative phase of the force and velocity oscillations, it is clear that the angle, ϕ, by which $\hat{v}_2$ lags $\hat{F}_a$ is, from inspection of Fig. 7.9 equal to $\theta_m - \pi/2$ so that, from Eqn. (7.17)

$$\phi = \cos^{-1} (F_s/\hat{F}_a) \qquad (7.21)$$

Now Eqns. (7.20) and (7.21) permit the calculation of the peak value of the actual (or total) v_2 wave-form and its phase displacement from the peak of the applied force wave-form F_a whereas it is the amplitude and phase shift of the fundamental component of $v_2(t)$ which is strictly required to determine the describing function $G_F(\hat{F}_a, s)$. As already mentioned, however, the describing-function approach is itself approximate through its neglect of harmonics and it is therefore reasonable to regard the vector $(\hat{v}_2/\hat{F}_a)$, $\angle\phi$ as a close approximation to the desired transfer function $\{G_F(\hat{F}_a, j\omega)\}/\{j\omega M\}$. We therefore have that

$$G_F(\hat{F}_a, j\omega) \simeq (\hat{v}_2 \omega M/\hat{F}_a), \angle(\phi - \pi/2) \qquad (7.22)$$

the modulus being obtainable from Eqn. (7.20) and the argument from Eqn. (7.21). From these equations it is immediately obvious that neither of these two coordinates is frequency dependent, the right-hand sides being functions only of amplitude $\hat{F}_a$ and the friction parameters F_s and F_b so permitting the argument $j\omega$ to be dropped from the function G_F, i.e.

$$G_F = G_F(\hat{F}_a) \qquad (7.23)$$

For given F_s and F_b values, therefore, a single locus of G_F, with $\hat{F}_a$ as parameter, may be calculated and drawn in polar form, but only for the range of $\hat{F}_a$ for which discontinuous oscillation occurs, i.e. for which Eqn. (7.12) holds. Substituting Eqns. (7.14) and (7.16) yields the upper bound for $\hat{F}_a$, given by

$$\hat{F}_a/F_b < \{1 + (\pi\alpha/2)^2\}^{0.5} \qquad (7.24)$$

where

$$\alpha = F_s/F_b \qquad (7.25)$$

Above this bound, F_a produces a continuous response as sketched in Fig. 7.10, the motion and the frictional force changing sign at angles θ_0 and $\theta_0 + \pi$ from the commencement of the force cycle. Integrating the equations

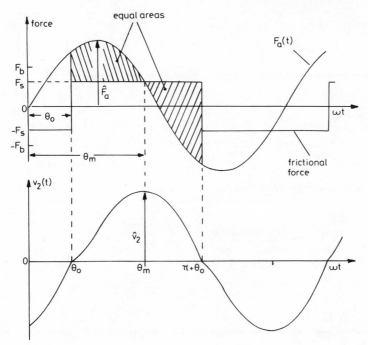

Fig. 7.10 *Continuous oscillation*

of motion between these limits gives

$$\theta_0 = \cos^{-1}\{(F_s/\hat{F}_a)(\pi/2)\} \tag{7.26}$$

the necessary condition that $\hat{F}_a/F_s > \pi/2$ being always met since motion will occur in practice if

$$\hat{F}_a/F_s < \{(\pi/2)^2 + (1/\alpha)^2\}^{0.5} \quad (> \pi/2)$$

as may be derived from constraint (7.24).

The expressions for θ_m and ϕ [Eqns. (7.17) and (7.21)] remain unchanged for continuous motion so that the phase of $G_F(\hat{F}_a)$ is also unaltered. The peak velocity $\hat{v}_2$ is now governed by an expression different from that of Eqn. (7.20), however, since on integrating the equation of motion we now get

$$\hat{v}_2 \omega M = \hat{F}_a(\cos\theta_0 - \cos\theta) + F_s(\theta_0 - \theta)$$

and on eliminating θ_0, using (7.26) and substituting for θ_m using Eqn. (7.17), we get

$$\hat{v}_2 \omega M/\hat{F}_a = \{1 - (F_s/\hat{F}_a)^2\}^{0.5} + (F_s/\hat{F}_a)[\cos^{-1}\{(F_s/\hat{F}_a)(\pi/2)\} \\ + \sin^{-1}(F_s/\hat{F}_a) - \pi/2] \tag{7.27}$$

and, under these conditions Eqn. (7.27) is the one from which $|G_F(\hat{F}_a)|$ should be calculated.

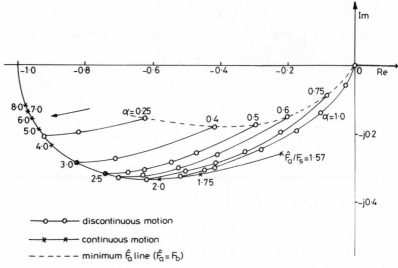

Fig. 7.11 *Locus of* $-G_F(F_a, j\omega)$

Figure 7.11 illustrates the resulting form of the loci of $-G_F(\hat{F}_a)$ which clearly changes for different ratios, α, of sliding to static friction as shown. The lower boundary of the loci is, however, independent of α as would be expected since this portion of the locus derived from Eqn. (7.27), represents continuous motion which is obviously unaffected by F_b. The calibration of the individual locus branches is given in Table 7.1.

Table 7.1 *Values of* $\hat{F}_a/F_s$ *pertaining to points plotted in Fig. 7.11*

α	Point							
	1	2	3	4	5	6	7	8
0·25	4·00	4·12	4·20					
0·40	2·50	2·75	2·95					
0·50	2·00	2·10	2·20	2·30	2·40	2·50		
0·60	1·67	1·83	2·00	2·15	2·28			
0·75	1·33	1·47	1·60	1·73	1·87	2·05		
1·00	1·00	1·11	1·25	1·33	1·50	1·70	1·75	1·86

7.2.5.2 Prediction of chatter. Between the two points A and B marked on the system block diagram of Fig. 7.5 there are effectively two transfer functions arranged in a negative feedback configuration as illustrated in Fig. 7.12. The forward path transfer function is the frequency-independent describing function $G_F(\hat{F}_a)$ already determined, whilst the feedback transfer function

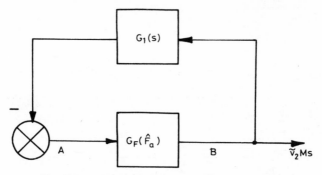

Fig. 7.12 *Block diagram reduced to a single loop*

$G_1(s)$ is derived by reducing the network of series and parallel transfer functions connected around the two points A and B, and describing the remainder of the system. Limit cycling will clearly occur at a frequency ω and applied-force amplitude $\hat{F}_a$ at which

$$G_1(j\omega)G_F(\hat{F}_a) = -1\cdot0$$

i.e. when

$$G_1^{-1}(j\omega) = -G_F(\hat{F}_a) \tag{7.28}$$

where the inverse Nyquist locus of $G_1(s)$ intersects the locus of $-G_F(\hat{F}_a)$. The advantage of having isolated $G_F(\hat{F}_a)$ in its present form is that the effect of varying all the system parameters (other than F_s and F_b) may be investigated by redrawing only the locus of $G_1^{-1}(s)$, $G_F(\hat{F}_a)$ remaining unchanged.

With the automatic control loop open, $G_1^{-1}(s)$ is readily shown to be given by

$$G_1^{-1}(s) = Ms^2/[k_a + k_h\{1 - \exp(-T_ds)\}] \tag{7.29}$$

the locus of $G_1^{-1}(j\omega)$ clearly oscillating in phase between the limits $\pi \pm \gamma$, where

$$\gamma = \sin^{-1}\{k_h/(k_a + k_h)\} \tag{7.30}$$

and crossing the real axis at distances x, where

$$x = -\frac{(n\pi)^2 M}{T_d^2(2k_h + k_a)} \quad n = 1, 3, 5$$

$$x = -\frac{(n\pi)^2 M}{T_d^2 k_a} \quad n = 0, 2, 4 \tag{7.31}$$

at frequencies

$$\omega = n\pi/T_d \quad n = 1, 2, 3 \tag{7.32}$$

as illustrated in Fig. 7.13.

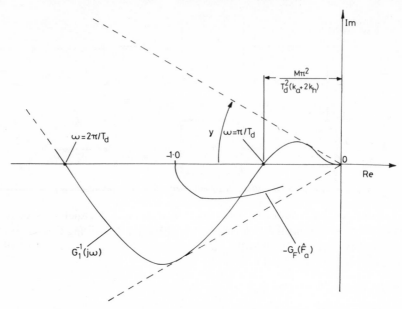

Fig. 7.13   *General form of* $G_1^{-1}(j\omega)$ *locus*

The ratio k_a/k_h of the *stiffnesses* of the transmission and cutting loops is clearly an important parameter in dictating the frequency of any limit cycling which must, from observation of the loci $G_1^{-1}(j\omega)$ and $-G_F(\hat{F}_a)$ for the system of Section 7.2.4 shown in Fig. 7.14, clearly occur in the range

$$\pi T_d^{-1} < \omega < 2\pi T_d^{-1} \tag{7.33}$$

for $k_a/k_h \ll 1 \cdot 0$† for machines of large mass M: as we have already observed in the simulation of this system. Values of ω, $\hat{F}_a$ and $\hat{v}_2$ determined from the intersection of $G_1^{-1}(j\omega)$ and $-G_F(\hat{F}_a)$ are also found to accord well with simulation. Increasing the stiffness of transmission above k_h clearly begins to reduce γ and compress the oscillation of the $G_1^{-1}(j\omega)$ locus about the negative real axis by moving the frequency of intersection towards the limiting value $\sqrt{k_a/M}$ so that chatter is then dictated by transmission resonance rather than cutting-head dynamics. This prediction is likewise confirmed by simulation.

The incorporation of the control-loop dynamics into $G_1(s)$ has but a minor effect on the locus shape in the vicinity of its intersection with $-G_F(\hat{F}_a)$, provided the gain K is relatively small so that control action can do little under such circumstances to combat high-frequency chatter. The low-frequency behaviour of the automatic control loop is considered in the following section of this chapter.

---

† The minor role played by the transmission under these circumstances is confirmed by the small proportion of ripple exhibited by tension, τ, in Fig. 7.6.

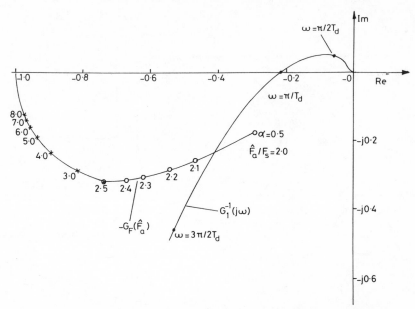

Fig. 7.14 *Intersection of* $-G_F$ *and* G^{-1} *loci for system of Section 7.2.4*

7.2.5.3 Prediction of low-frequency behaviour. For the purposes of low-frequency servo design the dynamics of the cutting loop (1) in Fig. 7.5 may be drastically simplified. Static and sliding frictional effects are ignored altogether making $G_F = 1{\cdot}0$ and the regenerative cutting dynamics

$$\tilde{y}(s)/\tilde{v}_2(s) = \{1 - \exp{(-T_d s)}\}/s \qquad (7.34)$$

[obtained by taking Laplace transforms of Eqn. (7.3)] approximated by expanding the right-hand side of Eqn. (7.34) and ignoring powers of s above the first so that

$$\tilde{y}(s)/\tilde{v}(s) \simeq T_d/(1 + 0{\cdot}5T_d s) \qquad |T_d s| \ll 1{\cdot}0 \qquad (7.35)$$

As an alternative approach, recalling that the use of Eqn. (7.3) is itself an approximation, the cutting dynamics may be modelled, for low-frequency effects only, by imagining the coal or mineral buttock wall to move forward not in discrete slices but at a continuous velocity $v_b(t)$ as it is attacked by the drum moving with a linear velocity $v_2(t)$. Figure 7.15 illustrates this concept. Only in steady state will v_b and v_2 reach equality and under transient conditions the differential equation

$$dy(t)/dt = v_2(t) - v_b(t) \qquad (7.36)$$

will apply. Now $v_b(t)$ is the rate at which material is torn from the buttock and therefore

$$v_b(t) = y(t)/T_d \qquad (7.37)$$

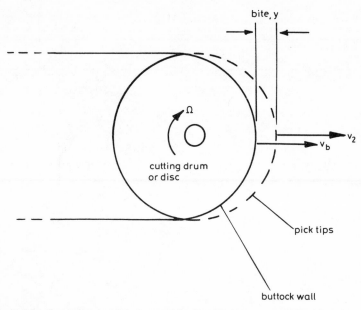

Fig. 7.15 *Conceptual physical model for simplified treatment of cutting dynamics*

so that elimination of $v_b(t)$ between Eqns. (7.36) and (7.37) produces the result

$$\tilde{y}(s)/\tilde{v}_2(s) \simeq T_d/(1 + T_d s) \tag{7.38}$$

The two approaches clearly produce first-order lag models for the cutting dynamics, but have time constants of different magnitude. These are fortunately significantly smaller than the dominant system lags so that the resulting differences in the predicted behaviour of the complete system are only marginal.

Ignoring friction, Eqn. (7.6) becomes

$$\tau(t) = F_h(t) + M \, dv_2(t)/dt = k_h y(t) + M \, dv_2(t)/dt \tag{7.39}$$

and with this simplification and that of, say, Eqn. (7.35) the transfer function of the open control loop becomes

$$G(s) = \frac{K k_c k_d k_e \, T_d}{\begin{array}{c} s\{1 + T(1 - 0{\cdot}5T_d s)s + (s^2/\omega_0^2)\} \\ (1 + 0{\cdot}5T_d s)(1 + T_m s)(1 + T_e s)(1 + T_h s) \end{array}} \tag{7.40}$$

where

$$T = T_d k_h/k_a \tag{7.41}$$

and

$$\omega_0 = (k_a/M)^{0{\cdot}5} \tag{7.42}$$

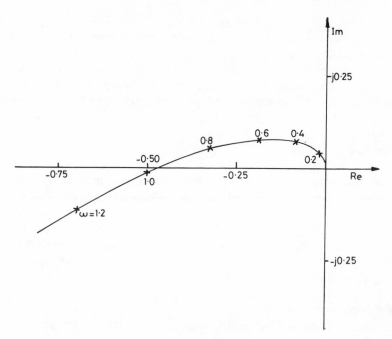

Fig. 7.16   *Inverse Nyquist locus for low-frequency analysis*

Now for highly elastic transmission $T \gg T_d$ and provided the resulting model predictions are confined to frequencies much less than $1/(0 \cdot 5 T_d)$ and ω_0 (equal to 5 and 4·93 rad/s in the example considered) $G(s)$ may be simplified to a cascade of first-order lags, thus

$$G(s) \simeq \frac{K k_c k_d k_e T_d}{s(1 + Ts)(1 + 0 \cdot 5 T_d s)(1 + T_e s)(1 + T_h s)(1 + T_m s)} \tag{7.43}$$

$$= 12 \cdot 4/\{s(1 + 2 \cdot 17 s)(1 + 0 \cdot 2 s)(1 + 0 \cdot 1 s)^2 (1 + 0 \cdot 05 s)\} \tag{7.44}$$

in the particular example considered earlier.

The open-loop $v_2(t)$ trace of Fig. 7.6 clearly shows a near exponential envelope having a time constant very similar to the 2·17 s predicted by Eqn. (7.44). The inverse Nyquist locus of $G(s)$ shown in Fig. 7.16 indicates a critical frequency of 0·97 rad/s (0·154 Hz) which is consistent with the simulated and observed frequencies of servo oscillation (0·120 and 0·128 Hz respectively) shown in Figs. 7.7 and 7.8. Had a lag of time constant T_d (=0·4 s), rather than $0 \cdot 5 T_d$ been used in the calculation of $G(s)$ then agreement would have been even closer suggesting that the continuous cutting-action concept is perhaps marginally superior to the discrete-bite concept for modelling of the cutting process in the low-frequency domain. Such a low-frequency model was indeed employed[7] long before the incorporation of detailed cutting-loop effects for the design of stable load controllers which eliminate the low-frequency oscillations exhibited by Figs. 7.7 and 7.8. The value of the detailed model lies, of course, in its potential to aid the elimination of chatter either by adjustments to machine design or by responsive regulator action.

7.3 Steering dynamics

We now turn attention from the process of cutting, as such, to the associated and equally important process of steering the cutting machine. Whilst cutting dynamics affect predominantly the rate of mineral extraction, steering dynamics affect the composition of the product if, in the interests of productivity, the size of the cutting head is close to the seam thickness. Good steering is, in any event, an essential prerequisite to safe and effective mining, yet its satisfactory control is far from simple.

7.3.1 Fundamental features
A winning machine will not follow automatically the seam or vein of desired material unless means are provided

(a) to deflect the direction of the cutting head in the two directions perpendicular to its line of travel
(b) to detect any deviation of the cutting head outside the boundaries of the desired material and into the unwanted surrounding strata
(c) to interlink operations (a) and (b) above by a suitable control strategy.

As in most physical steering processes operating in a terrestrial, marine or aerial environment, the steering mechanism, (a), effects a *change* in the *angle* of attack of the machine itself. At their simplest, therefore, the steering dynamics involve two cascaded integrations† between the deflection of the steering actuator (generally hydraulic in mining applications) and the resulting position of the machine in space. Furthermore, the error-sensing operation, (b), usually involves some transport delay because of the obvious engineering difficulty (if not impossibility) of siting a sensor at the point of cutting. Human observation of the position of the cutting head is restricted likewise because of the dense dust cloud created in its immediate vicinity.

Sensor and actuator dynamics are often significant in steering control, particularly the former because of the random nature of the low-power radiation techniques,[8, 9] often employed to detect the difference in strata characteristics. In some instances the alternative technique of monitoring the variation in cutting force is employed as a means of detecting movement of the cutting

---

† Integrations with respect to distance traversed in the general direction of cutting.

head outside the desired stratum but, because of the very random nature of the measurements, considerable delay and filter dynamics may be involved in producing the desired positional information.

These factors constitute the fundamental ingredients of the steering dynamics of tunnelling-type systems, i.e. those mining systems in which the direction of machine travel, whilst cutting, and the general direction of mineral extraction are colinear. The modelling and control problems for steering such systems, whilst being far from trivial, are comparatively straightforward. However, with longwall mining (which is by far the most common method of underground coal extraction in Europe—and increasingly popular in North America), serious additional complications are introduced requiring very careful consideration. These are now examined.

7.3.2 Geometrical modelling of a longwall system

Figure 7.17 shows diagrammatically those features and variables of a modern longwall shearer machine (and the associated mechanical system) essential to the modelling of its vertical steering characteristics. The machine makes repeated sweeps, or passes, of length L along the coal face, extracting on each pass a volume of material $= LW_dD$, where W_d and D are the drum width and diameter respectively, assuming perfect horizontal steering.† As indicated, the machine rides on the semiflexible structure of an armoured face conveyor (a.f.c.) of the scraper-chain type, and the cutting drum may be ranged up and down continuously in an attempt to keep the drum within the boundaries of the undulating coal seam.

Also as shown in Fig. 7.17, in a process normally termed the *pushover*, the a.f.c. structure is snaked forward horizontally onto the newly cut floor between passes, its front edge taking up a profile, in the vertical plane, similar to that of the cut floor alongside the new face. In this way, on pushover, the a.f.c. undergoes a tilt change at any point along its length (and in the face-advance direction), dependent upon the deflection J previously applied to the steering boom in the vicinity of the point in question. The tilt change results from the fact that the width W_c of the conveyor trays, or pans, is invariably greater than that of the drum, i.e.

$$W_c > W_d$$

Figure 7.18 illustrates in end elevation the tilting of the conveyor brought about by steering action.

The relationships of primary importance in modelling the steering system are basically geometrical involving variable heights and tilts which are, at first sight, functions of *two* independent variables—n, the pass number and l, the distance travelled along the face. In the interests of simplicity we shall con-

---

† That is assuming that the drum is fully embedded to a depth W_d in the coal face on each pass. Horizontal steering is briefly considered in Section 7.3.6.

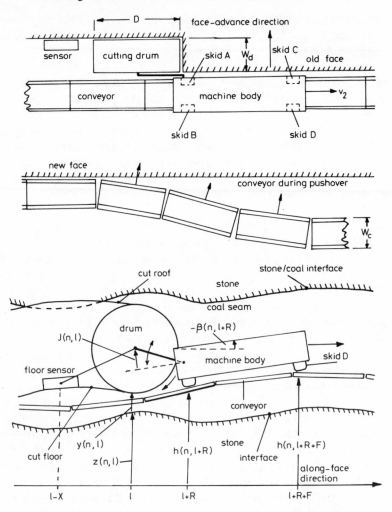

Fig. 7.17   *Diagramatic plan and side elevation of longwall shearer machine*

sider the special case of nearly equal drum and conveyor widths, viz.

$$W_c = W_d + \epsilon \tag{7.45}$$

with positive distance ϵ, whilst ensuring the tilt-change effect on pushover, being small enough to permit the approximation

$$W_c \simeq W_d = W \tag{7.46}$$

in developing the geometrical model. With this assumption, mere inspection of the side and end elevations of Figs. 7.17 and 7.18 reveals that, for small angular changes, the following relationships exist between the heights and tilts

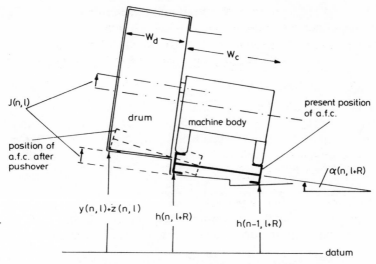

Fig. 7.18   *Diagramatic end view of longwall shearer system*

of the cut floor, seam and conveyor:

$$y(n, l) + z(n, l) = h(n, l + R) + W\alpha(n, l + R)$$
$$+ R\beta(n, l + R) + J(n, l) \tag{7.47}$$

$$\alpha(n, l) = \{h(n, l) - h(n - 1, l)\}/W \tag{7.48}$$

and

$$\beta(n, l) = \{h(n, l) - h(n, l + F)\}/F \tag{7.49}$$

where $y(n, l) + z(n, l)$ is the height of the cut floor at the new face wall, $h(n, l)$ that of the conveyor's front edge, $h(n - 1, l)$ that of the rear edge, $z(n, l)$ the height of the lower coal-stone interface and $J(n, l)$ the linear deflection of the steering boom from its null position. $\alpha(n, l)$ and $\beta(n, l)$ denote (in radians) the face-advance and along-face tilts of the machine respectively and are calculated on the assumption that the three skids A, B and C (Fig. 7.17) are in permanent contact with the a.f.c. The system parameters R and F denote the fixed offset distances between the drum and rear skids and between front and rear skids respectively. Equations of somewhat greater complexity could clearly be developed for the more general situation of unequal drum and conveyor widths.

Equations (7.47) to (7.49), in conjunction with some control law for the manipulation of $J(n, l)$, would permit the simulation of a single pass of the machine given profiles for $z(n, l)$, $h(n, l)$ and $h(n - 1, l)$, but the multipass solution for $y(n, l)$, $n = 1, 2, 3, \ldots$, etc., would require, in addition, a model of the pushover and conveyor-fitting process relating the next conveyor heights

$h(n + 1, l)$ and $h(n, l)$ to the newly produced floor profile $y(n, l) + z(n, l)$ and the previous floor $y(n - 1, l) + z(n - 1, l)$. The simplest form of model, here termed the *rubber conveyor* model, assumes complete flexibility of the structure such that, again assuming $W_c \simeq W_d$,

$$h(n + 1, l) = y(n, l) + z(n, l) \tag{7.50}$$

Such a model is *idealized* in the sense that it is the simplest model possible and reflects the conveyor designer's objective, namely flexibility, ideally. Unfortunately, as we shall see, the system behaviour, as predicted by this model, is far from ideal in the conventional sense of the word. The modelling of the a.f.c.'s partial rigidity is considered in Sections 7.3.4 and 7.3.5.

7.3.3 Simulated performance of system with rubber conveyor model
Using conventional analogue two-term feedback control based on measurements of machine tilt and delayed floor-coal thickness† according to the law

$$J_d(n, l) = -k_h y_m(n, l) - k_g W\alpha(n, l + R) \tag{7.51}$$

where k_h is the *height-gain* and k_g the *tilt- (or derivative height-) gain* of the controller and J_d is the demanded boom deflection; then, in presence of first-order sensor and actuator dynamics of *distance-constant‡* X_1 and X_2 respectively, viz.

$$\partial y_m(n, l)/\partial l = (1/X_1)\{y(n, l - X) - y_m(n, l)\} \tag{7.52}$$

and

$$\partial J(n, l)/\partial l = (1/X_2)\{J_d(n, l) - J(n, l)\} \tag{7.53}$$

a response typified by Fig. 7.19 is predicted by model equations (7.47) to (7.53). The predicted system behaviour is clearly totally unstable. This particular result is produced by a system having the parameters $k_h = 0.8$, $k_g = 1.0$, $X = 1.25$ m, $X_1 = 0.6$ m, $X_2 = 0.165$ m and $R = 0$. The setting of $R = 0$ implies a so-called *fixed-drum* machine in which there is no relative vertical movement between the machine body and drum, which is located above the rear skids, steering being accomplished by raising and lowering of the whole machine body in a pitching or sometimes rolling motion relative to the a.f.c. Making $R > 0$ only worsens the stability problem and no adjustment of the system parameters in the above model will achieve stability short of making

---

† In practice, roof- rather than floor-coal thickness is generally the more convenient measurement and the more crucial variable to control since roof-stone penetration can lead to serious roof collapse in many instances. With seams of nearly constant thickness the mathematical problem is little affected, but double drum machines, one for roof control and one for floor control, may be used in the event of a widely varying seam thickness (or in very thick seams) so posing a two-input, two-output modelling and control problem. This has been examined by Edwards and Greenberg.[11]

‡ Assuming a constant machine speed v_2, the time constants of the coal sensor and steering actuator are readily converted to the distance base of the steering problem.

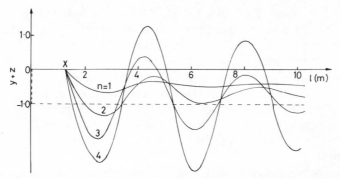

Fig. 7.19 *Simulated response of rubber conveyor model to unit downward step in coal seam*

$X = 0$ which has proved physically impossible with conventional radiation transducers. Stability can, however, be achieved by adopting a control strategy based on coal thickness data collected on the previous pass and stored within a computer for nearly one pass length, $L - X$, before being applied to the controller. Equation (7.52) is thus replaced by

$$\partial y_m(n, l)/\partial l = (1/X_1)\{y(n - 1, l) - y_m(n, l)\} \tag{7.54}$$

This predicted need to develop computer control—an enormously expensive undertaking for the underground environment—clearly makes worthwhile a much more detailed examination of the a.f.c.'s response to an undulating cut floor, in the hope that its limited flexibility might provide a sufficiently powerful stabilizing influence to permit the use of cheap analogue control. The problem is as yet unsolved because the physics involved is complex and the consequences of error due to oversimplification would be enormously costly. A number of simplistic approaches have been tried yielding inconclusive results but a very rational approach outlined in the next section holds promise of a reliable solution.

7.3.4 Modelling of the semiflexible conveyor
As indicated in Figs. 7.17 and 7.18, the structure of the a.f.c. comprises a chain of steel trays loosely joined end to end. The conveyor is semirigid in several senses, viz.

(a) the side channels of each tray are very stiff
(b) the deck plate of each tray is fairly stiff
(c) free angular movement between successive trays is hard—limited—in the along-face direction to $\pm \Delta \gamma$
(d) in the face-advance direction to $\pm \Delta \alpha$

where $\Delta \gamma$ and $\Delta \alpha$ are constants. Over and above these considerations are the facts that the joints will exhibit some elasticity and the cut floor may undergo some degradation (e.g. planing-off of high spots and filling of hollows with fine material) during the pushover phase. Furthermore, being fairly light compared to the machine weight, the a.f.c. structure may not remain completely stationary as the machine passes over.

Because of this complexity there exists a strong temptation to fit some simplified model of preassumed structure, perhaps using carefully scaled physical models[12] in the first instance for parameter estimation. Such a mathematical model is suggested in Section 7.3.5.3 but for the moment we attempt the rigorous inclusion of at least some of the above-mentioned factors in a derived (rather than a guessed) model and assess its potential for enhancement.

To demonstrate the principle of the method with minimal complication we shall in fact include only factors (*a*) and (*c*) above and disregard (*b*), (*d*) and the secondary effects initially. The deck plates are therefore imagined to remain completely flexible (as in the rubber conveyor concept) whilst the side channels are rigidized. The problem therefore reduces to fitting two independent chains of loosely joined stiff *rods* to two separate undulating *line* profiles beneath them. Each chain and line may therefore be treated independently. Figure 7.20 illustrates the problem variables involved, the along-face axis

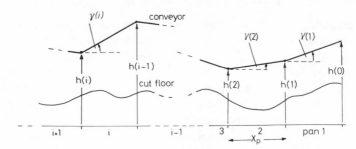

Fig. 7.20 *Definition of variables for conveyor fitting problem*

having been subdivided into I pan lengths, X_p, and $h(i)$ denoting the height of the left-hand end of rod i counted from the right-hand end of the face. In terms of the earlier notation, therefore, but dropping the pass number n

$$h(i) \equiv h(L - iX_p) \qquad i = 0, 1, 2, \ldots, I \tag{7.55}$$

where

$$I = L/X_p \tag{7.56}$$

Assuming small angles, the heights of successive joints are clearly related thus

$$h(i - 1) = h(i) + X_p \gamma(i) \tag{7.57}$$

and intermediate heights are given by

$$h(l) = h(i) + (l - L + iX_p)\gamma(i) \qquad L - iX_p < l < L - (i - 1)X_p \tag{7.58}$$

Now in the absence of elastic bending, the a.f.c. will settle on the undulating cut floor such that its total potential energy E_I is a minimum subject to the two constraints

$$h(l) \geq y(l) \qquad 0 < l < L \tag{7.59}$$

and

$$|\gamma(i) - \gamma(i - 1)| \leq \Delta\gamma \qquad i = 2, 3, \ldots, I \tag{7.60}$$

The potential energy δE_i of rod i is given by

$$\delta E_i = mg\{h(i) + h(i - 1)\}/2 \qquad i = 1, 2, \ldots, I \tag{7.61}$$

where $2mg$ is the weight of the pan, or from Eqns. (7.61) and (7.57)

$$\delta E_i = mg\{h(i) + X_p\gamma(i)/2\} \qquad i = 1, 2, \ldots, I \tag{7.62}$$

and the total potential energy of pans 1 to i inclusive is obviously

$$E_i = \sum_{j=1}^{i} \delta E_j$$

Now we may regard $h(i)$ as a *state variable* and $\gamma(i)$ as an adjustable *control variable* in an optimal control problem requiring the minimization of E_I with respect to the sequence of controls $\gamma(1)$, $\gamma(2)$, ..., $\gamma(I)$, subject to the above conveyor equations and constraints. The modelling problem has thus been conceptually transformed into an optimal control problem† to which the principle of general dynamic programming[13] is admirably suited for its solution. This principle translates the original multistage decision process (in this case the minimization of E_i with respect to the sequence $\gamma(1)$, $\gamma(2)$, ..., $\gamma(i)$) into i single-stage decision processes: in this case the minimization of E_i with respect to $\gamma(i)$ only, given the *minimum* total energy E^*_{i-1} of pans 1 to $i - 1$ as a function of $h(i - 1)$. For this application the principle may be expressed thus

$$E_i^*\{h(i)\} = \min_{\gamma(1),\, \gamma(2),\, \ldots\, \gamma(i)} [E_i\{h(i),\, \gamma(1),\, \ldots,\, \gamma(i)\}]$$

$$= \min_{\gamma(i)} [\delta E_i\{h(i),\, \gamma(i)\} + E^*_{i-1}\{h(i - 1)\}] \tag{7.63}$$

Figure 7.21 flowcharts the computational procedure implied by Eqn. (7.63). Basically, having chosen some arbitrary initial height $h(i)$, then $\gamma(i)$ is incre-

---

† Most physical modelling problems can usually be viewed as energy minimizing problems to which the techniques of optimal control analysis are potentially applicable. For simple lumped parameter systems such an 'energy approach' is generally more laborious than the conventional use of force, momentum, flow balance concepts however.

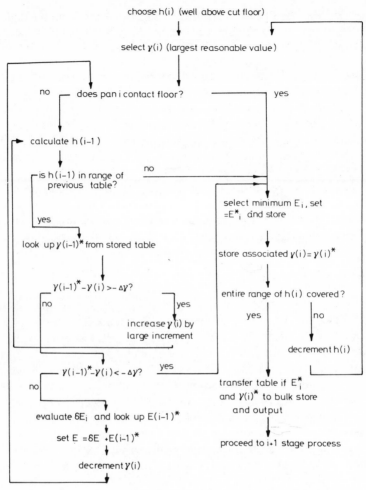

choose h(i) (well above cut floor)

select γ(i) (largest reasonable value)

no ─ does pan i contact floor?   yes

calculate h (i–1)

is h(i–1) in range of   no
previous table?

select minimum E$_i$, set
=E*$_i$ and store

yes

look up γ(i–1)* from stored table

store associated γ(i)= γ(i)*

γ(i–1)* – γ(i) > – Δγ?

entire range of h(i) covered?

no   yes

yes   yes   no

increase γ(i) by
large increment

decrement h(i)

γ(i–1)* – γ(i) < – Δγ?   yes

no

transfer table if E*$_i$
and γ(i)* to bulk store
and output

evaluate δE$_i$ and look up E(i–1)*

set E =δE +E(i–1)*

proceed to i+1 stage process

decrement γ(i)

Fig. 7.21   *Flow chart for fitting i th pan by dynamic programming*

mented downwards in small stages until one of the constraints (7.59) or (7.60) is impinged. With each trial value of $\gamma(i)$, δE_i is evaluated using Eqn. (7.62) along with $h(i-1)$ from (7.57) and hence $E^*_{i-1}\{h(i-1)\}$ may be obtained from a stored look-up table previously computed during the optimization of the $i-1$ stage process. The minimum value of $\delta E_i + E^*_{i-1}$, and associated $\gamma(i)$ are selected, set equal to E^*_i and $\gamma(i)^*$ respectively and stored alongside $h(i)$ whereupon $h(i)$ is incremented and the procedure repeated until tables of $E^*_i\{h(i)\}$ and $\gamma(i)^*$ covering the entire field of interest have been computed. Attention is now transferred to the $i+1$ stage process and minimization of E_{i+1} accomplished making use of the E^*_i table just obtained. At the conclusion of the exercise a sequence of energy minimizing angles $\gamma^*(1)$, $\gamma^*(2)$, ...,

$\gamma^*(I)$ is obtained for any given $h(I)$ from which the desired piecewise linear profile for the a.f.c. is obtained using Eqn. (7.57). Starting the entire procedure (i.e. with i set to unity) poses no difficulty in that all the entries in the E_0^* table are zero.†

The computational sequences described above clearly need not be and in fact are not prohibitive in terms of their execution time and storage requirements if attention is restricted to, say, 20 trays with a height resolution of perhaps 0·5% of the peak undulation amplitude. The reason is that this problem involves only one state—and one control—variable so that the well known 'curse of dimensionality' associated with higher order dynamic programming problems does not arise here. Stiffening of the deck plate [i.e. inclusion of factor (b)] and incorporation of the relative tilt constraint in the direction of face-advance [factors (d)] introduces some, but not excessive, complication. The state order remains unaltered but we are now faced with two control sequences $\gamma(i)$ and $\alpha(i)$ $(i = 1, 2, ..., I)$ so that, associated with each E_i^* there will also exist a floating range of tilts

$$\alpha_1^*(i) < \alpha(i)^* < \alpha_2^*(i)$$

determined by incrementing $\alpha(i)$, for each chosen $\gamma(i)$, to find those values which cause either (i) contact with front and rear floors, or (ii) constraint $|\alpha(i) - \alpha(i - 1)| \leq \Delta\alpha$ to be impinged.

An additional inner computational loop for varying $\alpha(i)$ is therefore involved and the constraint-testing procedures are now slightly more complicated.

If elasticity at the joints is permitted then the energy to be minimized becomes the total potential-plus-strain-energy so that if E_i now represents the *total* energy per pan

$$\delta E_i = mg\{2h(i) + X_p \gamma(i)\} + k_d\{\delta\gamma(i)\}^2 + k_b\{\delta\alpha(i)\}^2 \qquad (7.64)$$

where the absolute elastic yields $\delta\gamma(i)$ and $\delta\alpha(i)$ are given by

$$\delta\gamma(i) = |\gamma(i) - \gamma(i - 1)| - \Delta\gamma$$
$$\delta\alpha(i) = |\alpha(i) - \alpha(i - 1)| - \Delta\alpha \qquad (7.65)$$

and the stiffness coefficients k_a and k_b governed by

$$k_a = \text{constant} \quad \delta\gamma(i) > 0$$
$$\quad = \text{zero} \qquad \delta\gamma(i) < 0$$
$$k_b = \text{constant} \quad \delta\alpha(i) > 0$$
$$\quad = \text{zero} \qquad \delta\alpha(i) < 0$$

† The foregoing discussion has presupposed free end conditions on $h(0)$ and $h(I)$. The computational sequence for $i = 1$ and $i = I$ would differ somewhat from the general procedure of Fig. 7.21 if the ends were fixed.

Joint elasticity therefore involves only slight complication of the cost function and minimal increase in program execution time.

The dynamic programming approach therefore holds great promise for a soundly based rational solution to the problem of conveyor simulation. The a.f.c. fitting routine is obviously run between simulations of the along-face dynamics described by Eqns. (7.47) to (7.49) and (7.51) to (7.53).

7.3.5 Modelling for analytical predictions
Simulation of a complicated system is, of course, fraught with pitfalls and, even when properly running, can be laborious and time consuming if there exists no prior idea of the broad behavioural characteristics and the useful range of parameters to be explored. Analytic solutions, albeit of perhaps simplified systems, can be of enormous assistance in overcoming these difficulties, but, for many years, it was not appreciated how a multipass system might be rendered amenable to solution by the control engineer's conventional techniques.

7.3.5.1 The use of long transport delays to model the interaction between passes. A breakthrough leading to a more familiar form of model for multipass systems finally occurred when it was realized that the spatial coordinates n and l are not independent variables in the sense that, say, time and distance are independent in the more conventional distributed parameter systems (see Volume I, Chapter 2). In particular n and l are *not independent of one another* since l varies continuously but repeatedly in the range $0 < l < L$, integer n being incremented by a unit step with each resetting of l. This means that any point specified by the two coordinates n, l could alternatively be specified by the *single* coordinate $v =$ total distance cut (or passed) by the machine. In an *unidirectional* cutting scheme such as that illustrated in Fig. 7.22 in which the machine cuts only from left to right,† v is clearly given by

$$v = l + nL \qquad (7.66)$$

The result of this realization is that the dependent variables, say $y(n, l)$ or $h(n, l)$, at a point n, l may be expressed alternatively as functions of v only, i.e. $y(v)$ and $h(v)$ respectively in this case. Extending the argument, $h(n - 1, l)$ could, for example, be expressed in the new coordinate basis as $h(v - L)$. Process equations (7.47) to (7.49) may therefore be rewritten in the new basis as follows:

$$y(v) + z(v) = h(v + R) + W\alpha(v + R) + R\beta(v + R) + J(v) \qquad (7.67)$$

---

† Bidirectional schemes in which the machine cuts alternately in each direction are also amenable to analysis[10] in terms of v but in this situation sampled-data system concepts must also be introduced.

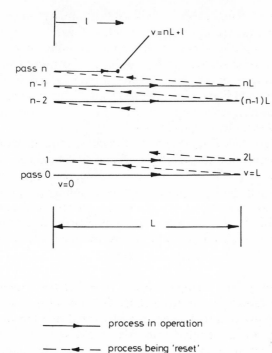

pass n

n-1

n-2

1

pass 0

v=0

v=nL+l

nL

(n-1)L

2L

v=L

L

———► process in operation

— —◄ — process being 'reset'

Fig. 7.22 *Showing the accumulation of total distance passed, v, in a unidirectional multipass process*

where

$$\alpha(v) = \{h(v) - h(v - L)\}/W \tag{7.68}$$

and

$$\beta(v) = \{h(v) - h(v + F)\}/F \tag{7.69}$$

Additionally Eqn. (7.50) describing the rubber conveyor could be re-expressed in the form

$$h(v + L) = y(v) + z(v)$$

so that

$$h(v) = y(v - L) + z(v - L) \tag{7.70}$$

Likewise the control law (7.51) and Eqns. (7.52) and (7.53) describing the sensor and actuator dynamics may also be expressed in terms of v rather than n and l giving

$$J_d(v) = -k_h y(v) - k_g W\alpha(v + R) \tag{7.71}$$

$$dy_m(v)/dv = (1/X_1)\{y(v - X) - y_m(v)\} \tag{7.72}$$

and

$$dJ(v)/dv = (1/X_2)\{J_d(v) - J(v)\} \tag{7.73}$$

By taking Laplace transforms in s with respect to v of Eqns. (7.67) to (7.73), a block diagram of transfer functions involving lags X_1 and X_2 and transport delays X and L could clearly be obtained and manipulated to produce an open-loop system transfer function which could be analyzed for closed-loop system stability.

As an example we shall ignore, for simplicity of presentation, the sensor and actuator dynamics so that Eqns. (7.71) and (7.72) reduce to

$$y_m(v) = y(v - X) \tag{7.74}$$

and

$$J(v) = J_d(v) \tag{7.75}$$

so that the control law (7.71) becomes simply

$$J(v) = -k_h y(v - X) - k_g W\alpha(v + R) \tag{7.76}$$

Again for simplicity we set $k_g = 1.0$ so that effect of the tilt α on the process behaviour is nullified and attention is restricted to the special case of the fixed-drum shearer so that offset R may be set to zero. Combining process Eqns. (7.67) and (7.70) with control law (7.76) for this special case gives

$$y(v) + z(v) = y(v - L) + z(v - L) + J'(v) \tag{7.77}$$

where

$$J'(v) = -k_h y(v - X) \tag{7.78}$$

so that in terms of Laplace transforms taken with respect to v, we obtain

$$\tilde{y}(s) + \tilde{z}(s) = \{\tilde{y}(s) + \tilde{z}(s)\} \exp(-Ls) + \tilde{J}'(s) \tag{7.79}$$

where

$$\tilde{J}'(s) = -k_h \tilde{y}(s) \exp(-Xs) \tag{7.80}$$

yielding the block diagram given in Fig. 7.23. From the block diagram we see that the system may be regarded as a unity feedback system having the inverse open-loop transfer function

$$G^{-1}(s) = k_h^{-1} \exp(Xs)\{1 - \exp(-Ls)\} \tag{7.81}$$

Figure 7.24 gives the form of the inverse Nyquist locus derived from Eqn. (7.81) for the case of $X = L/12$, the closed-loop system being clearly unstable for all k_h since the inverse Nyquist stability criterion[10, 14] requires that the net number of counterclockwise encirclements by $G^{-1}(s)$ of the point $-1 + j0$ in the plane of $G^{-1}(s)$, as s describes the usual clockwise semicircular

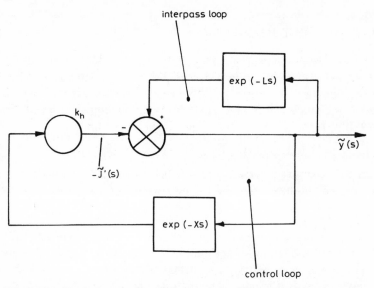

Fig. 7.23 *Block diagram representation of simplified vertical steering system*

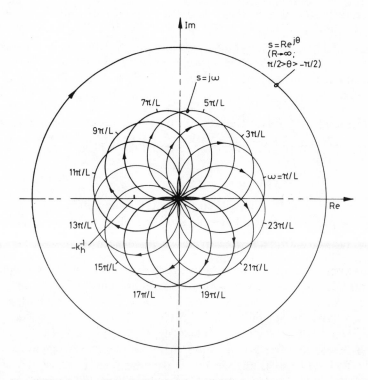

Fig. 7.24 *Inverse Nyquist diagram for simplified vertical steering system*

excursion around the entire right-hand half-s-plane, should equal the number of zeros of $G(s)$ ($=0$ in this case) within that half-plane.

7.3.5.2 Implicit approximations. The prediction of closed-loop instability by the delay model Eqn. (7.81) accords completely with simulation experience and predictions are unaltered if the sensor and actuator dynamics are included in the analytical model, so confirming the simulation result given in Fig. 7.19. If X is set $= L$ then from Eqn. (7.81) we see that

$$G^{-1}(s) = k_h^{-1}\{\exp{(Ls)} - 1\} \tag{7.82}$$

and from the resulting inverse Nyquist locus it is readily deduced that stability can now be obtained provided the positive gain k_h is restricted such that

$$k_h < 2{\cdot}0 \tag{7.83}$$

which again confirms the results of simulation studies reported in Section 7.3.3. Indeed the modelling of the multipass process in terms of v and the consequent introduction of transport delays of duration L has proved to be a most reliable analytical means of predicting their behavioural characteristics generally. This is provided that (*a*) the pass length L is always very much greater than the settling time of the transients occurring along the pass and (*b*) attention is confined to transients induced at considerable distances from both ends of the pass.

These limitations are important so far as mathematical rigour is concerned and highlight the fact that the process equations (7.67) to (7.73) and all subsequent equations derived from them are in fact valid only for values of v in the ranges

$$nL + X < v \le (n + 1)L \qquad n = 0, 1, 2, \ldots \tag{7.84}$$

since the sensor cannot commence measurement until $l > X$. For values of v within the ranges

$$nL \le v \le nL + X \qquad n = 0, 1, 2, \ldots \tag{7.85}$$

the process is repeatedly reinitialized (i.e. reset) by boundary-condition equations not considered hitherto but which might perhaps take the form

$$y(v) = h(v) = J(v) = y_m(v) = 0 \qquad nL \le v \le nL + X \tag{7.86}$$

if the machine were assumed to be properly 'launched' at the start of each pass. This resetting action of course represents a series of discontinuities in the process occurring at intervals L which are completely disregarded in the continuous-delay model presented in Section 7.3.5.1 and, furthermore, this resetting would occur, albeit instantaneously, even if $X = nL$ ($n = 0, 1, 2$ etc.). The delay model is therefore not suited to situations in which it allows transients to flow through from one pass to the next or in which the real-life boundary conditions themselves induce transients. Owens[15] has presented an alternative view of the stability of multipass systems using the techniques of functional analysis and has discussed the rigour of the delay model as a basis for stability prediction.

7.3.5.3 Approximate representation of conveyor 'dynamics'. In Section 7.3.4 we considered carefully the modelling of the conveyor for computer simulation. This exercise yielded insight into the causes of conveyor stiffening but, without undertaking an exhaustive simulation programme, the model provides little insight into the likely *effects* on overall system behaviour. The model might reasonably be criticized on the grounds that it obscures the wood by the trees. Taking the broader view it is intuitively reasonable to expect that the introduction of some rigidity into the conveyor model might exercise a damping influence on the steering process because of the a.f.c.'s inability to follow precisely the shape of the cut floor beneath. Such an argument can, however, lead to a paradox because to work at all, the steering process relies on the flexibility of the a.f.c. Qualitative arguments cannot therefore settle the question as to whether or not the real-life steering system should be stable, or for what range of parameters stability might be expected. Approximate analytical models, if logically constructed, should, however, be capable of indicating the degree of damping which might be anticipated from detailed simulations so justifying expensive simulation studies of showing them to be unworthwhile. It is the development of such analytical models (capable of integration into the delay model of the multipass process) which we now consider.

If the rubber conveyor equation (7.70) were modified to the slightly more general form

$$h(v) = k_a\{y(v - L) + z(v - L)\} \tag{7.87}$$

where k_a is an attenuating constant in the range

$$0 < k_a \leq 1 \cdot 0 \tag{7.88}$$

then the transfer function appearing in the interpass loop of Fig. 7.23 would be modified from $\exp(-Ls)$ to $k_a \exp(-Ls)$ and the inverse open-loop transfer function $G^{-1}(s)$ would become

$$G^{-1}(s) = k_h^{-1} \exp(Xs)\{1 - k_a \exp(-Ls)\} \tag{7.89}$$

so modifying the inverse Nyquist locus to the form shown in Fig. 7.25. This indicates that stability could now be achieved with positive controller gains k_h, provided

$$k_h < 1 - k_a \tag{7.90}$$

Attenuation of interpass coupling would therefore appear to be beneficial but the inclusion of merely the constant k_a takes no account of the fact that the

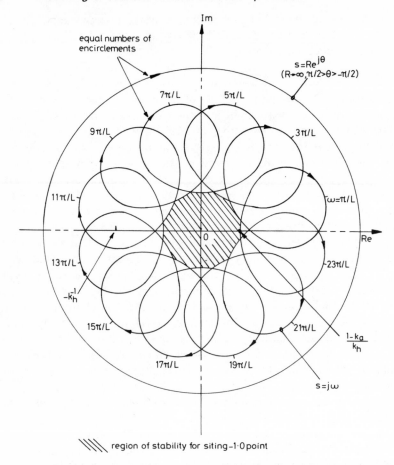

Fig. 7.25   *Inverse Nyquist locus with constant interpass attenuation*

a.f.c. will respond differently to different frequencies and also to different amplitudes.

Figure 7.26 illustrateś the general shape which one edge of the a.f.c. would adopt (if unrestrained by the deck plate) in response to a sharp upward undulation in the cut floor, (*a*) beneath a pan joint, and (*b*) beneath a pan centre point. The illustration also shows curves closely fitting the true piece-wise linear profiles and which could provide the basis for modelling the a.f.c. by linear dynamics. Clearly a first-order linear system is capable of generating an impulse response resembling case (*a*) and perhaps an overdamped second-order system might generate that shown for case (*b*). Both cases exhibit fairly similar rates of decay, however, so that the two possible linear models would not differ drastically in their basic dynamics. The unique feature common to

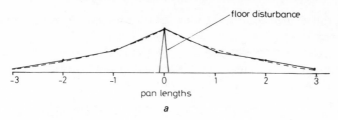

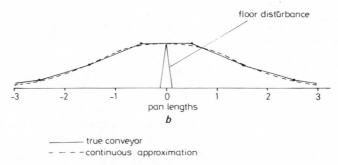

———— true conveyor
— — — continuous approximation

Fig. 7.26 *Response of conveyor to a sharp disturbance in the cut floor*
a Disturbance between pans
b Disturbance at pan-centre

both impulse responses is, however, their symmetrical two-sided† nature. Such behaviour is not usually encountered in the study of physical dynamical systems but can occur in the field of off-line data processing.

Only bidirectional dynamical systems, i.e. those which process the input signal in both forward *and backward* time sequence, are capable of producing such two-sided responses. A system to produce the approximate impulse response shown in Fig. 7.26(*a*), for instance, would operate on the input signal as follows:

(i) the input data stream is stored
(ii) this data is now injected through a filter of transfer function $0{\cdot}5/(1 + X_c s)$, where X_c is the distance constant of decay, and the output stored
(iii) the original input data is fed from store in reverse-time sequence through an identical filter to that above and the output stored

† Obviously, if the floor impulse were off-centre the response would still be two-sided though no longer symmetrical. Floor disturbances will occur in practice with equal probability to left and right of centre, however, so justifying the use of a symmetrical impulse response approximation to the true response of the a.f.c.

(iv) the time sequence of the output data from step (iii) is reversed and the result stored
(v) the stored responses obtained from steps (ii) and (iv) are added and stored
(vi) the stored output from step (v) is fed out of the system after a total delay L from the commencement of the input signal.

Such a system cannot, of course, operate in real time unless it is associated with a long transport delay (the interpass delay L in this case). The frequency transfer function for the above process, provided

$$L \gg X_c \tag{7.91}$$

is given by

$$H_c(j\omega) = \{0 \cdot 5/(1 + X_c j\omega) + 0 \cdot 5/(1 - X_c j\omega)\} \exp(-Lj\omega)$$
$$= \exp(-Lj\omega)/\{1 + (X_c \omega)^2\} \tag{7.92}$$

which would now replace the rubber conveyor transfer function $\{= \exp(-Lj\omega)\}$ in the interpass loop of Fig. 7.23. Again, for the validity of the model, any disturbances considered must occur well away from the ends of the pass to prevent the a.f.c. response from impinging on the process boundaries: a fuller mathematical discussion of this and related questions has been published by Edwards.[11]

Whereas $G^{-1}(s)$ is unaffected[11] as Laplace variable s sweeps around the positive half-s-plane at infinite radius, for $s = j\omega$ we now have that

$$G^{-1}(j\omega) = k_h^{-1} \exp(Xj\omega)\{1 - H_c(j\omega)\}$$
$$= k_h^{-1} \exp(Xj\omega)[1 - \exp(-Lj\omega)/\{1 + (X_c \omega)^2\}] \tag{7.93}$$

so that as ω increases towards the value $1/X_c$ the attenuation effect of the a.f.c. rigidity becomes effective reducing the size of the lobes of the inverse Nyquist locus in the manner shown in Fig. 7.27 and opening up an area of stability for the siting of the critical point.

Stability may apparently be possible, therefore, if positive gain k_h is restricted such that

$$k_h < 1 - 1/\{1 + (\pi X_c/X)^2\} \tag{7.94}$$

so placing the $-1 + j0$ point within the area of stability. The result therefore suggests that stability is potentially possible but there exists considerable uncertainty as to the effective value for X_c. This value will of course increase with signal amplitude indicating the possibility of limit cycling. All predictions made on the basis of this type of analytical model must be regarded as tentative, however, because of the very considerable approximations made in its original formulation.

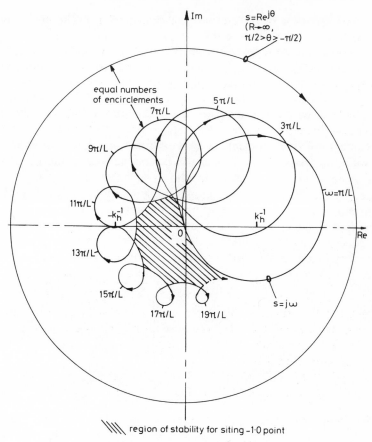

Fig. 7.27   *Inverse Nyquist locus with first-order a.f.c. model*

7.3.6 Other multipass processes

Other multipass processes are encountered in industry and generally involve the processing of some material or workpiece by repeated passes of the processing tool. In some cases the tool is stationary and the workpiece moves whilst in others the reverse situation applies. Examples include (i) the rolling and (ii) the machining of metals. Example (ii) has much in common with (iii) the horizontal (rather than vertical) steering of coal-face machines since high spots produced on pass n can cause the partial rejection of the cutting tool on pass $n + 1$ because of the deeper penetration required, so possibly accentuating the original roughness. In these examples the interpass interaction comes about through the yield of the tool-positioning structure and should be small in well-designed systems. The interaction between passes is therefore usually a comparatively weak secondary effect unlike the vertical steering system analyzed previously in which such interaction is the primary effect. In applications (i) and (ii), however, and in the case of thin web machines—(iii)—good surface finish can be a very important requirement so that the weak interpass effects must nevertheless be considered.

Another example of a related type is the automatic agricultural tractor which is steered to follow the previously ploughed furrow. The prediction of electrical load demand on power systems and the associated control of generation might also be viewed as a multipass process because of the strong statistical correlations which exist between demand patterns (*a*) *along* the day in question and (*b*) *across* consecutive days.

The simulation of multimachine systems such as vehicle convoys, multistand rolling and paper mills etc. can also be conducted as a multipass process since a chain of identical transfer-function matrices (T.F.M.'s) $G(s)$ may be simulated by the repeated excitation of a single simulation cell of T.F.M. $G(s)$ from the stored (i.e. delayed) output from the previous simulation. The machines are thus simulated sequentially rather than simultaneously.

The concept may be extended as Edwards[16] has shown to the simulation of linear spatially distributed systems approximated in the manner of Chapter 2, Volume 1, by a chain of similar discrete simulation cells.

The stability and dynamic response of all such systems may therefore be investigated using delay models of the type developed in the previous section. Although presented here in a specifically mining context it is hoped that through this discussion the reader will appreciate the relevance of the suggested modelling approach to a much wider class of problem. The analogy with two-dimensional data-processing systems has also been mentioned in Section 7.3.5.3 and the wealth of theoretical analysis[17, 18] published in this area may also be usefully brought to bear on those multipass processes of a more physical type.

7.4 Clearance systems

We now turn attention from the winning faces towards the mine shaft or drift up which the coal or mineral is conveyed to the surface for preparation for market. The system for transporting the material between these two points is known as the clearance system and constitutes a network of conveyors and fixed intermediate bunkers for smoothing out bottlenecks. The conveyors may be continuous (i.e. belt conveyors or discrete) for example shuttle cars or trains of tubs or mine cars hauled by winches or locomotives. Discrete types of conveyor or variable-speed belts may also serve temporarily as bunkers and some bunkers can, as we shall see in Section 7.4.2, also resemble conveyors to some extent rather than being simple hoppers. The distinction between the purely storage function and the purely transportation function can therefore become somewhat blurred in practice.

The individual entities of the clearance system can pose interesting local control problems demanding very detailed modelling of the sub process itself, and an example of this is considered in Section 7.4.2. A somewhat less detailed but wider ranging model of the clearance network itself, or a substantial portion of it, is also necessary as an aid to the synthesis of central (computer) control systems for the supervision of the overall production from the mine. The question of decomposition of a complex network into handlable units naturally arises and the provision of a large bunker can often serve to define the boundary between such units. Underground bunkers are, however, extremely expensive to install because of the cost of the large excavations needed for their accommodation and it is therefore worthwhile to consider whether or not sophisticated flow control of a more comprehensive sort might not obviate the need for an enormous bunker. By implication, therefore, the size of the bunker may be left unspecified at the control-design stage and the bunker cost be included as one term of an overall cost function to be minimized. Having determined the integrated optimal control strategy for the entire network the necessary optimal bunker sizes can then be determined by simulation. (The alternative approach more commonly practised is to simulate, on a cut-and-try basis, arbitrarily sized bunkers and control systems of a simple localized type to determine the best of the guessed combinations, none of which are likely to provide a globally optimal performance.)

7.4.1 Modelling clearance systems for optimal control synthesis

Figure 7.28 illustrates an elementary clearance system comprising two winning faces (a) and (b) supplying mineral to a common drift via trunk conveyors involving transport delays T_a and T_b respectively. We shall assume that a constant volumetric rate $f_c(t)$ is demanded by the drift conveyor, any excess delivery from the trunk conveyors being accommodated by a pit-bottom

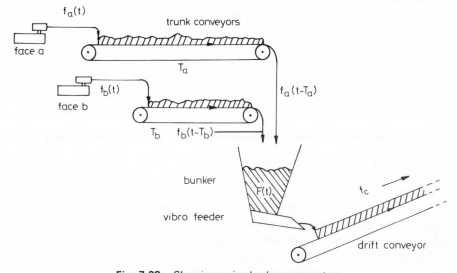

Fig. 7.28 *Showing a simple clearance system*

bunker as shown and any deficiency made up from the bunker. If $f_a(t)$ and $f_b(t)$ denote the instantaneous production rates from the winning machines (a) and (b) respectively, then if $F(t)$ is the volumetric contents of the bunker it is obvious that

$$dF(t)/dt = f_a(t - T_a) + f_b(t - T_b) - f_c(t) \qquad (7.95)$$

the variables being subject to the overriding constraints that

$$F(t), f_a(t), f_b(t), f_c(t) > 0 \qquad (7.96)$$

Now the machine speeds and, hence, $f_a(t)$ and $f_b(t)$ are controllable for much of the time but production is halted cyclically during face-end manoeuvres and also randomly particularly during periods of unfavourable geological conditions at the face when machine progress may be temporarily halted by roof fall or damage to equipment. The face-production rates may therefore be described by the equations

$$f_a(t) = d_a(l_a, t)u_a(t)$$
$$f_b(t) = d_b(l_b, t)u_b(t) \qquad (7.97)$$

where $u_a(t)$ and $u_b(t)$ are the demanded production rates adjusted by manipulation, when possible, of the machine haulage speeds $v_a(t)$ and $v_b(t)$ (no distinction between the speeds of the two ends of the haulage transmission is necessary here because the haulage system dynamics discussed in Section 7.2 are far faster than those of the clearance system). The variables $d_a(l_a, t)$ and $d_b(l_b, t)$ switch between the values of unity and zero, partly randomly in time for reasons mentioned above and partly cyclically in distance l_a and l_b at face-length intervals L_a and L_b. The along-face distances are of course governed by the differential equations

$$dl_a(t)/dt = v_a(t) = f_a(t)/W_a$$
$$dl_b(t)/dt = v_b(t) = f_b(t)/W_b \qquad (7.98)$$

(where W_a and W_b denote the web thicknesses) and the constraints

$$0 < l_a < L_a$$
$$0 < l_b < L_b \qquad (7.99)$$

The statistical properties of face-stoppages have been exhaustively studied in the U.K. and European coal mining industries and incorporated into a number of computer simulation packages[19, 20, 21] for the study of clearance system behaviour. For control system synthesis, however, the random stoppage signals (d_a and d_b in this case) have to date only been modelled[21, 22] by white noise fed into low-order colouring filters whose time constants match the mean time between stoppages and mean stopping duration. The cyclic component could no doubt be incorporated fairly readily by employing a second-order oscillatory model of appropriate natural frequency though the distance-base might pose problems.

The prime costs to be balanced in the optimization of a coal clearance system are the costs of production which are basically flow-rate dependent, and cost of bunkerage which are dependent on stockpile sizes. If F_c denotes bunker capacity, therefore, the performance index to be minimized might reasonably take the form, in our example,

$$C(f_a, f_b, T_f) = \int_0^{T_f} [\lambda_1\{f_a^i(t) + f_b^i(t)\} + \lambda_2 F_c^j] \, dt$$

$$= \lambda_2 \int_0^{T_f} [F_c^j + \lambda\{f_a^i(t) + f_b^i(t)\}] \, dt \qquad (7.100)$$

where

$$\lambda = \lambda_1/\lambda_2 \qquad (7.101)$$

λ_1 being the total annual cost of the winning and conveying machinery (p.u. flow rate^i) including electricity costs, interest, depreciation and maintenance costs etc., and λ_2 the total annual cost of bunkerage (p.u. stored-volume^j). The parameters λ_1 and λ_2 may be fitted after a careful study of available cost data. T_f is the period over which the process is to be optimized, this being at least the duration of a production shift and longer in the case of continuous (3-shift) production. In the interests of ease of solution of the optimal control problem, i.e. that of minimizing C by choice of input functions $u_a(t)$ and $u_b(t)$ it is tempting to choose indices $i = j = 2$ but in practice i should probably be of greater magnitude and j nearer unity since bunker costs are likely to depend more-or-less linearly on F_c whereas the penalty for overloading electrical machines, in terms of reduced life expectancy and the consequential losses of production through insulation failure, tends to occur more abruptly with increasing load than the quadratic index would suggest.

The block diagram for the system of Fig. 7.28 is shown in Fig. 7.29 but for optimal control calculations it is preferable to relegate the transport delays to the output end of the diagram as shown in Fig. 7.30. Provided $T_f \gg T_a$ or T_b as in this application, this step results ultimately in a control law demanding feedback from only the ends of the conveyor delays and not from intermediate points along the conveyors—as Noton[13] has demonstrated and we shall see shortly in a simple example. The optimization of large networks is most conveniently accomplished by the use of discrete dynamic programming techniques[13, 21, 22] necessitating the modelling of the conveyors T_a and T_b by n and m discrete slices respectively as indicated in Fig. 7.30 where

$$n = T_a/\Delta T \quad \text{and} \quad m = T_b/\Delta T \qquad (7.102)$$

ΔT being the time step for the computations and selected to be considerably smaller than T_a or T_b. (The model now resembles a continuous train of

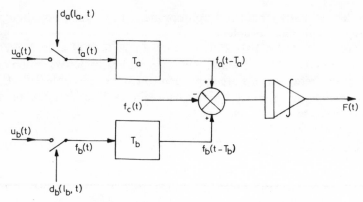

Fig. 7.29 *Block diagram of clearance system of Fig. 7.28*

fixed-speed mine cars with variable filling.) A state transition matrix model of the form

$$\mathbf{x}\{(i + 1)\,\Delta T\} = \boldsymbol{\Phi}(\Delta T)\mathbf{x}(i\,\Delta T) + \boldsymbol{\Delta}(\Delta T)\mathbf{u}(i\,\Delta T) \tag{7.103}$$

where $\mathbf{x}$ is the state vector and $\mathbf{u}$ the input vector ($= [u_a, u_b]^T$) is now formed from the n equations of delay T_a, viz.

$$x_{a,1}\{(i + 1)\,\Delta T\} = x_a(i\,\Delta T)$$
$$x_{a,j}\{(i + 1)\,\Delta T\} = x_{a,j-1}(i\,\Delta T) \qquad 1 < j \leq n \tag{7.104}$$

and m similar equations in $x_{b,j}$ for delay T_b, together with the $2p$ state equations describing the dynamics of the pth order colouring filters describing d_a and d_b and the state equations

$$x_a\{(i + 1)\,\Delta T\} = x_a(i\,\Delta T) + \Delta T\,f_a(i\,\Delta T)$$
$$x_b\{(i + 1)\,\Delta T\} = x_b(i\,\Delta T) + \Delta T\,f_b(i\,\Delta T) \tag{7.105}$$

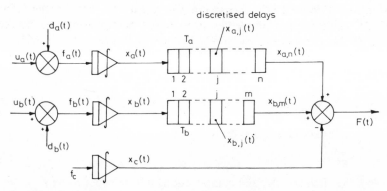

Fig. 7.30 *Block diagram of clearance system modified for optimal control design*

$$x_c\{(i + 1)\, \Delta T\} = x_c(i\, \Delta T) + \Delta T\, f_c(i\, \Delta T) \tag{7.106}$$

$$f_c\{(i + 1)\, \Delta T\} = f_c(i\, \Delta T) \tag{7.107}$$

d_a and d_b are now approximated to additive rather than modulating disturbances so that

$$f_a(t) \simeq d_a(t) + u_a(t) \quad \text{and} \quad f_b(t) \simeq d_b(t) + u_b(t) \tag{7.108}$$

The state vector x is therefore of order $n + m + 2p + 4$ and the process model is clearly obtainable in the linear form of state-transition equation (7.103).

As regards the cost model [Eqn. (7.100)], if i and $j = 2 \cdot 0$ we have the convenient integral quadratic function of state and input variables, provided bunker capacity F_c can be linearly related to $F(t)$. Ideally of course F_c should represent the upper limit of $F(t)$ but provided occasional spillage and conveyor shut-down can be tolerated then F_c might be equated to the root-mean-square value of $F(t)$ so allowing a reasonable margin for filling above the mean level $\bar{F}$ and only a low probability of emptying the bunker. We therefore define F_c as follows:

$$F_c^2 = \lim_{T_f \to \infty} \int_0^{T_f} \frac{F^2(t)}{T_f}\, dt \tag{7.109}$$

so permitting the replacement of F_c^2 by $F^2(t)$ in Eqn. (7.100). In simulating the optimized system it is found, however, that $\bar{F}$ settles at a negative value so contravening constraint (7.96). The problem is readily overcome by noting that

$$\lim_{T_f \to \infty} \int_0^{T_f} \frac{F^2(t)}{T_f}\, dt = \lim_{T_f \to \infty} \int_0^{T_f} \left|\frac{2\bar{F} - F(t)}{T_f}\right|^2 dt = \bar{F}^2 + \sigma_F^2 \tag{7.110}$$

where σ_F denotes the standard deviation of $\bar{F} - F(t)$ and, hence, substituting $\{2\bar{F} - F(t)\}^2$ for F_c^2 in the performance index. The fact that $\bar{F}$ is the unknown a priori is unimportant, its value being determined in the course of simulation of the optimal control law. It is merely necessary to increase the reference signal $2\bar{F}$ progressively until the probability of emptying the bunker is reduced to an acceptably low level. The performance index in summation form therefore becomes, from Eqns. (7.100), (7.95), (7.104), (7.105), (7.106) and (7.110)

$$C = \lambda_2 \sum_{i=0}^{T_f/\Delta T} [\{2\bar{F} - x_{a,n}(i\, \Delta T) - x_{b,m}(i\, \Delta T) + x_c(i\, \Delta T)\}^2$$

$$+ \lambda\{f_a^2(i\, \Delta T) + f_b^2(i\, \Delta T)\}]\, \Delta T \tag{7.111}$$

The modelling of the clearance system in the manner indicated and the application of dynamic programming (a very efficient version[13] of which has been specially tailored to the optimization of linear processes subject to quadratic costs) has been found to yield very successful control strategies ideally suited to central computer control and which perform well dynamically and economically not only on the linearized plant models but also on the original non-linear system. Furthermore the resulting control rules are reducible to a very straightforward physical interpretation. This is now demonstrated with the aid of a simple clearance system example involving only one winning machine.

7.4.1.1 A simple example. Channel (b) of the winning and clearance system of Fig. 7.28 is here removed leaving only channel (a). In the foregoing equations, therefore, terms involving suffix (b) are set to zero and suffix (a), being now unnecessary, is dropped from the notation. This simple system is amenable to solution by pencil-and-paper methods using Pontryagin's maximum principle, as will be briefly outlined, and there is therefore no need in this particular case to discretize the process equation and cost integral. Our problem is therefore that of minimizing, by choice of $u(t)$, the cost function

$$C(t) = \lambda_2 \int_0^\infty [\{2\bar{F} - x(t - T) + x_c(t)\}^2 + \lambda f^2(t)] \, dt$$

where $f(t) = u(t) + d(t)$ and the state equations are

$$dx(t)/dt = u(t) + d(t) \tag{7.112}$$

and

$$dx_c(t)/dt = f_c = \text{constant} \tag{7.113}$$

Now considering the cost of bunkerage we note that

$$\int_0^\infty \{2\bar{F} - x(t - T) + x_c(t)\}^2 \, dt = \int_{-T}^\infty \{2\bar{F} - x(t) + x_c(t) + f_c T\} \, dt$$

$$+ \int_0^\infty \{2\bar{F} - x(t) + x_c(t) + f_c T\} \, dt$$

and since the first integral on the right-hand side (above) cannot be controlled by $u(t)$, $t > 0$ we confine attention to minimizing

$$C = \lambda_2 \int_0^\infty [\{2\bar{F} - x(t) + x_c(t) + f_c T\}^2 + \lambda\{u(t) + d(t)\}^2] \, dt$$

By the maximum principle of Pontryagin this implies maximizing the Hamiltonian

$$H = p(t)\{u(t) + d(t)\} - [\{2\bar{F} - x(t) + x_c(t) + f_c T\}^2 + \lambda\{u(t) + d(t)\}^2]$$

where costate $p(t)$ is governed by

$$dp(t)/dt = -\partial H/\partial x = 2\{x(t) - x_c(t) - f_c T - 2\bar{F}\} \tag{7.114}$$

For maximum H we set $\partial H/\partial u = 0$ giving

$$2\lambda\{u(t) + d(t)\} = p(t) \tag{7.115}$$

For the optimum control profile $u(t)$ we must therefore solve Eqns. (7.112) to (7.114) simultaneously in reverse time τ $(= T_f - t)$. As $\tau \to \infty$ we find that

$$u(\tau) + d - f_c \to -(1/\sqrt{\lambda})\{x(\tau) - x_c(\tau) - 2\bar{F} - f_c T\}$$
$$\to \{x(0) - x_c(0) + \sqrt{\lambda}f_c - 2\bar{F} - f_c T\} \exp{(\sqrt{\lambda}\tau)}/2 \tag{7.116}$$

and since $F(t) = x(t - T) - x_c(t)$, the control law (7.116) may be expressed

$$u(t) = -d(t) + f_c(t) - (1/\sqrt{\lambda})\{x(t) - x(t - T) + F(t)\}$$
$$+ (1/\sqrt{\lambda})\{2\bar{F} + f_c T\} \tag{7.117}$$

The mean level $\bar{F}$ therefore appears as a reference signal as expected to be set in simulation. The term $\{x(t) - x(t - T) + F(t)\}$ is very interesting, being the *total* mineral stored instantaneously in the bunker and on the conveying system. $F(t)$ may be measured directly and the conveyor storage term computed from a single-belt weigher together with a digital delay and integration routine. The control law is therefore highly practical, calling for feedback of 'total system storage'. Analysis of the control law by means of inverse Nyquist techniques[23] reveals it to be non-resonant, yet not overdamped, for a very wide range of cost ratios, λ.

This simplified example has not brought out the need for machine load-control with which we were preoccupied in Section 7.2 of this chapter. This is partly because of our neglect of the dependence of cost weighting, λ, on the variable mineral hardness. As Edwards has shown[22] however, load-control does appear if the thermal time constants of the production machines are modelled, and the cost of overtemperature penalized rather than merely the instantaneous production rates. The resulting control of the process is a good compromise between the conflicting requirements of the clearance system and the production machines, and takes advantage of the inherent overload capacity of electrical machines to sustain mine output during periods of interruption of supply from other faces.

7.4.2 The modelling of a horizontal bunker conveyor

Global network models of the type described in the previous section of this chapter are simplistic in that they take for granted the operation of local control loops for the manipulation of flow rates from feeders, levels in bunkers, cutting-machine speeds etc. These local loops can, however, involve unit subprocesses, the dynamic behaviour of which can be highly complex. This was evident in our study of coal-cutting dynamics earlier in the chapter. In the clearance system, too, individual equipments can pose local modelling and control problems which are far from trivial and which must be solved for the mine to operate in any fashion, optimally or otherwise. The horizontal

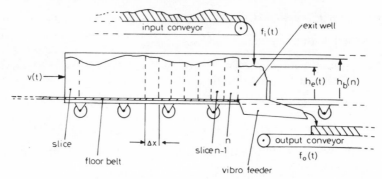

Fig. 7.31   *Diagram of bunker conveyor*

bunker conveyor is one such example which, as will be seen, operates in a manner which is far more complex than might be envisaged from a higher level in the control hierarchy where the bunker is viewed conceptually as a simple hopper.

Figure 7.31 illustrates such a bunker diagrammatically and indicates the main variables required for its mathematical description. Mineral enters the bunker randomly at a volumetric flow rate $f_i(t)$ directly into the exit well which supplies the vibrating feeder from which mineral is discharged to the output conveyor as shown. It will be assumed in this analysis that the output rate $f_0(t)$ can be independently manipulated although in practice this implies the existence of a tightly tuned control loop around the vibro feeder and incorporating a weigher on the output belt situated some distance downstream. This is generally essential to compensate for the effect of varying well height, $h_e(t)$, and mineral quality on $f_0(t)$. Even in this minor loop however, considerable modelling and control problems are posed.[24]

Any excess of inflow may be drawn into the bunker by setting velocity $v(t)$ to a negative value whilst a reduced inflow may be compensated by means of mineral tipped back into the exit well from the bunker, i.e. by making $v(t)$ positive. Delivery to and from the bunker tends to occur in discrete slices of horizontal thickness Δx (typically 0·6 m but depending on mean particle size and moisture content). In practice the slices tend to be separated by inclined shear planes but we shall assume these to be vertical for simplicity of analysis. The heights of the individual slices are variable in an uncontrolled situation and, as indicated in Fig. 7.31 are denoted by $h_b(i)$, where integer i takes the range of values

$$i = 1, 2, 3, \ldots, n \tag{7.118}$$

where n is the number of slices in the bunker at time t, being constrained such that

$$1 \leq n \leq N \tag{7.119}$$

where N is the total number of slices that the bunker can accommodate.

In modelling it is therefore assumed that, if W is the bunker width, each time, t_n, the bunker moves forward by a distance Δx a volume of mineral $W \Delta x\, h_b(n)$ is tipped into the exit well and the number n is decremented by 1·0. Each time the bunker travels backwards by distance Δx, however, a slice of height $h_e(t_n^-)$ is transferred to the bunker from the exit well and the well contents fall abruptly in volume by an amount $W \Delta x\, h_e(t_n^-)$, where t_n^- denotes the time marginally before the material transfer occurs. The derivative of $h_e(t)$ therefore changes impulsively with each material transfer so that, if A is the cross-sectional area of the well, the behaviour of $h_e(t)$ may be described by the differential equations

$$A\, dh_e\,(t)/dt = f_i(t) - f_0(t) + W \Delta x\, h_b(n)\, \delta(t - t_n) \qquad v(t_n^-) > 0 \quad (7.120)$$

or

$$A\, dh_e\,(t)/dt = f_i(t) - f_0(t) - W \Delta x\, h_e(t_n^-)\, \delta(t - t_n) \qquad v(t_n^-) < 0 \quad (7.121)$$

where

$$h_b(n) = h_e(t_n^-) \qquad v(t_n^-) < 0 \qquad\qquad\qquad\qquad (7.122)$$

and the transfer time t_n is given by

$$\int_0^{t_n} v(t)\, dt = n\, \Delta x \qquad\qquad\qquad\qquad\qquad (7.123)$$

$\delta(t - t_n)$ denotes a unit-impulse function occurring at $t = t_n$.

Real-time simulation of the system is highly desirable for operator training and for the investigation of automatic control schemes. Equations (7.120), (7.121) and (7.123) are readily simulated on an analogue computer equipped with elementary logic to detect incremental bunker movements exceeding Δx, the sign of $v(t_n^-)$, and to switch in the impulse function at the appropriate instants t_n. Simple analogue-control systems based on measurements of $h_e(t)$ are likewise amenable to analogue simulation. Since N is a large integer (100 to 200 for example) the storage of the bunker-height profile $h_b(i)$ ($i = 1, 2, \ldots, n$), however, requires the use of a digital computer interfaced to the analogue machine via appropriate converters. Elements are added to the array $h_b(i)$ whenever $v(t_n^-) < 0$ and withdrawn whenever $v(t_n^-) > 0$ in accordance with Eqns. (7.120), (7.121) and (7.122).

Figure 7.32 illustrates the behaviour of the bunker controlled according to control law (7.124), the purpose of which is to attempt to maintain the bunker level as close as possible to reference h_r. The control law is expressed thus

$$v(t) = k_p\{h_r - h_e(t)\} + k_i \int_0^t \{h_r - h_e(t)\}\, dt \qquad\qquad (7.124)$$

where k_p and k_i are the proportional and integral gains of the controller and Fig. 7.32 shows the system response when disturbed by step changes in $f_i(t)$ as

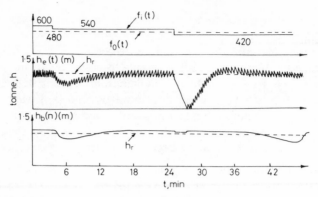

Fig. 7.32   *Response of automatically controlled bunker to input flow changes*

shown. The system parameters are $W = 2$ m, $A = 4$ m^2, $\Delta x = 0.4$ m, $h_r = 1.5$ m. The reduction in $f_i(t)$ from 600 to 540 tonne/h is clearly compensated, after a transient dip in h_e and h_b, by a reduction in absolute velocity, as evidenced by the reduced frequency at which slices enter the bunker. The fall of $f_i(t)$ to 420 tonne/h, i.e. below the constant output rate of 480 tonne/h, is also clearly compensated after a considerable transient by a reversal of the bunker role from filling to emptying. Interesting features of the process dynamics are:

(*a*) That transients created in the $h_b(i)$ profile whilst the bunker is filling will ultimately return to disturb the control system when the bunker is emptying.
(*b*) That whereas the uncontrolled process is of a fundamentally self-regulating nature during the filling operation [the heights $h_b(i)$ and $h_e(t)$ settling to some constant value for constant $v(t)$, $f_i(t)$ and $f_0(t)$], its dynamics change to an integrating type during emptying. The controller parameters for satisfactory filling are therefore not necessarily applicable whilst emptying and indeed stability in both modes is difficult to achieve with constant controller parameters.

Similar modelling techniques could readily be applied to other mineral-handling processes which share the same principle of operation such as the filling and emptying of trains of mine cars, for instance. In some instances the last car filled might be the first to be emptied, so involving the 'record-and-reverse' procedure used to simulate the conveyor bunker. In other cases, when filling and emptying take place in the same sequence, the storage process will resemble a transport delay, but will be physically discretized in a manner resembling our earlier mathematical segmentation of the continuous conveyor belt.

7.5 Discussion

The chapter has illustrated the novelty of the models of mining processes occurring throughout the mine from the winning face to the pit bottom. Despite the apparently special nature of mining processes, important links with processes outside the mining field have, however, been identified. We have, for example, noted similarities in the dynamics of mineral cutting with those occurring in the machining of metals. The multipass nature of longwall coal mining is a characteristic which has been shown to be shared by various repetitive processes, industrial and otherwise. The techniques for developing mining process models, whether for direct simulation, analytical solution or controller synthesis, can therefore be usefully applied outside the mining field and should therefore be of general interest.

Special characteristics which have appeared throughout the chapter have included long transport delays associated with a variety of phenomena including regenerative cutting dynamics, multipass operations and conveying. Their incorporation into models for simulation and for solution by frequency-response analysis and optimal control design techniques has been demonstrated, producing worthwhile control results and a good degree of insight into why the systems behave as they do. Also encountered has been the 'record and reverse' process—associated with bidirectional coal-face operation, the representation of face-conveyor dynamics and in modelling bunker and train dynamics. Its incorporation into simulation has been demonstrated and it can generally be incorporated into analytical studies although this has here been restricted to face-conveyor studies.

7.6 Acknowledgments

The modelling of a.f.c. behaviour by dynamic programming described in Section 7.3.4 was first undertaken by the author as part of a contract financed by a grant from the Mining Research & Development Establishment of the National Coal Board. The N.C.B.'s permission to publish the material of Section 7.3.4 in the present text is gratefully acknowledged. The views expressed are those of the author and not necessarily those of the N.C.B. It is also acknowledged that substantial portions of the remainder of the chapter were originally reported by the author in References 6, 7, 9 to 12, and 22 to 23, but with the emphasis there on stability and control rather than on process modelling as such.

7.7 References

1 WELBOURNE, D. B., and SMITH, J. D.: 'Machine tool dynamics' (Cambridge University Press, 1970)
2 MERRITT, H. E., and HOHN, R. E., 'Chatter, another control problem', *Control Eng.*, December, 1967, pp. 61–64
3 O'DOGHERTY, M. J., and BURNEY, A. C.: Colliery Engineering, Feb. 1963, **40**, pp. 51–54 and March, 1963, pp. 111–114
4 EVANS, I., and POMEROY, C. D.: 'Strength, fracture and workability of coal' (Pergamon Press, Oxford, 1966)
5 POMEROY, C. D.: 'Breakage of coal by wedge action—factors affecting tool design', *Colliery Guardian*, July, 1964, **209**, pp. 115–121
6 EDWARDS, J. B.: 'Modelling the dynamics of coal and mineral cutting', *Proc. Inst. Mech. E.*, 1978, **192**(30), pp. 359–370
7 DOWELL, J., and EDWARDS, J. B.: 'Development of the Bretby external haulage', *Min. Electr. Mech. Eng.*, January, 1968, **49**(567), pp. 3–17
8 COOPER, L. R.: 'Gamma-ray backscatter gauges for measuring coal thickness on mechanical coal faces', Proceedings of IEE International Conference on Industrial Measurement and Control by Radiation Techniques, Guildford, 1972, IEE Conf. Pub. No. 84, pp. 89–93
9 EDWARDS, J. B., and ADDISON, G. J.: 'The nucleonic coal sensor as an element of a control system for automatic steering of a coal cutter', ibid., pp. 20–29
10 EDWARDS, J. B.: 'Stability problems in the control of multipass processes', *Proc. IEE*, 1974, **121**(11), pp. 1425–1432
11 EDWARDS, J. B., and GREENBERG, J. M.: 'Longitudinal interactions in multipass processes', *Proc. IEE*, 1977, **124**(4), pp. 385–392
12 BOGDADI, W. A., and EDWARDS, J. B.: 'The automatic vertical steering of a longwall coal-cutting machine—an experimental investigation', *Proc. Inst. Mech. E.*, 1975, **189**(32/75), pp. 187–195
13 NOTON, A. R. M.: 'Variational methods in control engineering' (Pergamon Press, London, 122 pp.)
14 SHINNERS, S. M.: 'Control system design' (Wiley, New York, 1964, 523 pp.)
15 OWENS, D. H.: 'Stability of linear multipass processes', *Proc. IEE*, 1977, **124**(11), pp. 1079–1082
16 EDWARDS, J. B.: 'Wider application of multipass systems theory', Part 1: 'Multimachine and multicell systems' and Part 2: 'Controlled distributed processes', *Proc. IEE*, 1978, **125**(5), pp. 447–452 and 453–459
17 JURY, E. I.: 'Stability of multi-dimensional scalar and matrix polynomials', *Proc. IEEE*, 1978, **66**(9), pp. 1018–1047
18 BOLAND, F. M., and OWENS, D. H.: 'Linear multipass processes: a two-dimensional interpretation', *Proc. IEE*—to be published
19 MITCHELL, G. H., and LEE, J. R.: 'Digital simulation in operational research: Simulation applied to underground transport problems in collieries of the National Coal Board', Hollingdale, S. H. (ed.) (English Universities Press, London)
20 RANYARD, J. C.: 'Locomotive coal transport study', Chap. 4, Ph.D. Thesis, 1972, University of Lancaster, pp. 45–52
21 FAWCETT, W.: 'Simulation and control of coal production and transport', Ph.D. Thesis, University of Sheffield, 1975
22 EDWARDS, J. B., and MARSHALL, S. A.: 'Integrated plant and control system design for the operation of a mine at minimum cost', Proc. of I.F.A.C., International Symposium on Automatic control in mining mineral and metal processing, Sydney, Australia, August, 1973

23 EDWARDS, J. B.: 'Optimal control strategies for systems of multiple bunkers, conveyors and supply points', Proc. of I.E.E. Conference on Measurement and control in the handling and processing of materials, London, April, 1978, pp. 153–160
24 DOWELL, J., and PARK, A. H. D.: 'The design and performance of control systems for horizontal bunkers and bunker conveyors', ibid., pp. 113–120

Chapter 8

Manufacturing systems

T. R. Crossley

8.1 Introduction

This chapter comprises two principal parts: firstly, Section 8.2 contains a general introduction to the organization and problems encountered in manufacturing industry, and secondly, Section 8.3 contains part of a structured analysis model of a company engaged in batch manufacturing.

Thus, in Section 8.2, the concept of a manufacturing system is described in some detail and discussed in terms of the production efficiency and flexibility of such systems. The activities and communications generally found in an engineering manufacturing system are described, as are the principal types of production: these include the methods of job production, batch production, and flow-line production. A brief review of mechanization and automation is included, and the introduction of numerically controlled machines is described.

Finally, in Section 8.2, reference is made to the *off-line processing* of data and the *on-line control* of production processes, and information in Computer-Aided Manufacturing Systems is discussed, and integrated manufacturing systems are introduced.

The concept of functional modelling of systems through the use of a structured decomposition technique is described briefly in Section 8.3, and two types of such models are introduced: firstly, a 'functional model' which is concerned principally with *activities*, and secondly, an 'information model' concerned principally with *data*. This technique, known as Structured Analysis and Design Technique, SADT™,* has been developed by a commercial company, SofTech Inc., Waltham, Massachusetts, MA 02514, USA. The models presented in this chapter are based on this technique but are not

* SADT™ is a Registered Trade Mark of SofTech Inc.

intended to be rigorous, being based only on available published literature.[1-5] However, they do highlight the principal features of this system which is being used by a number of major organizations engaged in manufacturing.

8.2 Manufacturing industry

8.2.1 General

This chapter is intended to be of interest to those involved in the manufacturing industries, that is those concerned with the large-scale, organized production of goods. These industries may be divided into two broad groups: firstly, those involved in the production of materials and commodities, and secondly, those involved in the production of discrete items. By definition, manufacturing industry involves organized activities. The organization extends to all phases of the activity from the identification of a saleable product and a suitable market, through the actual manufacturing processes, to the sale of the product and the provision of after-sales services, and it involves many extremely complex relationships.

The tasks of creating and maintaining the organization and of coordinating all phases of the manufacturing activity are those of management. The method, structure and goals of management vary from one concern to another and vary due to changes in economic, social and political climates: the aim may be to maximize profits and ensure high returns on invested capital, to maximize output at any cost or to maintain a stable work load and work force in time of economic uncertainty. In general, however, it may be said that the aim of management in any manufacturing organization is to deliver goods of a proper quality on time and at an acceptable cost. In order to achieve this it must be ensured that personnel with the appropriate skills are available, along with the necessary tools and materials, in the right place at the right time and all activities must be planned and scheduled, then carefully monitored and controlled to ensure that specific deadlines are met.

8.2.2 Manufacturing systems

The management of an engineering concern strives to control the interplay between personnel, materials and machines with a view to *optimizing* the activity as a whole. Traditionally, this control has been determined by a largely empirical approach. In the early 1960s, however, the idea of scientific engineering was conceived and the techniques of systems engineering were applied to manufacturing for the first time.

A system is a collection of interrelated parts which act together in accordance with a set pattern. The set of procedures employed by management, the information networks and the equipment available within a company may be regarded as parts of a system responsible for turning raw inputs into material product outputs. This so-called *manufacturing system* is concerned with all elements of design, planning, control, machining, assembly and testing processes. It consists of a system in its own right made up of individual subsystems connected by paths for communication. Information and energy are exchanged or shared between the parts and this in turn implies that the parts change with time and the system is dynamic. This is particularly true of manufacturing systems: changes in economic policy and the rising costs of materials and labour together with the social trends producing shortages of skilled workers have forced changes in the way many companies approach the problems of manufacture. The purpose of a manufacturing system is to convert inputs into outputs. Its performance in achieving this may be evaluated in terms of two criteria; the *production efficiency* measured in terms of labour and machine productivity, rate of stock turnover, delivery performance, etc., and *flexibility*, which is a measure of the ability of the system to respond to changing demands and resources. Many companies, in particular long-established concerns operating in areas of rapid technological advance, are finding that their manufacturing systems which evolved from intimate man-machine relationships of the nineteenth century are no longer efficient in the face of increased competition or flexible enough to keep pace with changing markets and technology. Machine tools and personnel have been added to increase capacity, ageing equipment has been replaced by more productive plant and new skills have been developed; but very rarely have changes been made to the actual manufacturing system laid down when the manufacturing unit was established.

By adopting the systems engineering approach it has been found that manufacturing activities can be represented by *mathematical models* which permit the use of powerful tools for synthesis, analysis and optimization. This not only permits a greater understanding of increasingly more complex systems but also permits more rapid changes to existing systems and reduces development times.

8.2.3 *Activities and communications*
The activities within an engineering manufacturing system may be divided into five broad areas:

(1) product identification, specification and design
(2) production scheduling and forward planning
(3) production planning and control
(4) actual manufacturing processes
(5) inspection and quality control.

These areas are interconnected by routes for the flow of products and information. If the system is to function effectively it must have facilities and methods not only for controlling the physical flows but also for the timely generation, collection and communication of information, and many of the problems in manufacturing are caused by poor communications between areas of activity.

The traditional conveyor of both physical items and information is man. It is now held that the degree to which it is possible to optimize the performance of a manufacturing system is directly related to the degree to which communications within the system can operate without human intervention. Planning for optimization in manufacturing is therefore synonymous with automating the manufacturing system from the design concept to the finished part. The area of manufacturing which has attracted most attention to date is the most labour-intensive area, the actual production process, and it is in this area that the most significant advances in automation have been made.

8.2.4 Types of production

Before considering the extent to which the automation of production processes has developed, it is necessary to identify the different types of production processes employed in manufacturing engineering. There are three main types of production, namely, job production, batch production and flowline production. All three tend to be closely associated and may overlap in many circumstances.

Job production describes the method by which single articles are manufactured. All engineering concerns, whatever their nature, are involved at some time or other in job production be it the manufacture of small components required for maintenance of plant, the production of prototypes or tools, small jobbing contracts for other concerns or large-scale job-type production such as shipbuilding. The general characteristics of job production systems result from their general-purpose nature and the wide variety of work they must perform. Usually a wide range of general-purpose, versatile machinery and equipment is available together with a staff of highly skilled personnel and a permanent store of standard materials and components to permit the manufacture of as great a variety of work as possible at short notice. Because of the general nature of the equipment used, however, and the lack of time for detailed optimization of each job, job-production systems tend to be inefficient in terms of manpower and machine productivity. The fluctuating demands on a job-production system make it necessary for the system to be highly flexible and change rapidly to suit each particular job. This is usually made possible because individuals, or small teams, are given responsibility for parts, or the whole of the job from beginning to completion and, therefore, the communication problems caused by transfer of authority, information and goods from one section of the system to another can be avoided.

Batch production may be defined as the manufacture of a product in small or large batches or lots, by a series of operations, each operation being carried out on the whole batch before any subsequent operation is started. Batch production is by far the most common method of working in manufacturing industry and it is estimated that approximately 75% of all parts produced by the metalworking industries are produced in batches of less than fifty. Batch manufacture is almost universally accomplished by issuing components into manufacture on an 'operation' basis; that is, the work to be done is split down into separate operations involving perhaps 5 to 30 operations per part. Each of these operations may involve passing the part from one manufacturing process to another and, even if many of the operations are confined to one machine tool, changes in set up or position of the workpiece on the table are very frequent. These changes in set up mean that the machine tool is not cutting during this period and the overall efficiency may be as low as 15 to 20%. When the workpiece is passed from machine to machine, this situation gets much worse as, even with good organizations, it is rarely possible to manage a large machine shop in such a way that components spend less than one day between operations. To ensure that all components produced in a batch are identical, it is necessary to employ strict inspection procedures which add to the inefficiency of the process.

Batch manufacture has several major disadvantages caused by the delays and movements between operations. These communications problems include

(*a*) large amounts of work in progress develop which involve large capital investments

(*b*) large production storage areas and generous transport facilities are needed and a very effective planning and control system is needed to meet production deadlines

(*c*) comparatively long production periods are needed due to the time that each batch has to wait before proceeding from one operation to the next.

Batch production presents the greatest problems in manufacturing due to the combination of poor efficiency and communications with the need to maintain a high degree of flexibility to enable a continuously altering plan of work output to be applied.

Flowline production is the manufacture of a product by a continuous series of operations, each article going on to a succeeding operation as soon as available. Flowline methods are usually only applied when components are required in very large numbers over long periods of time. The manufacturing system tends to be very rigid, and depends heavily on large financial investments on capital equipment which is designed and arranged, with knowledge of the type of component to be produced, to operate at optimum efficiency.

8.2.5 Mechanization and automation

The word automation, which is generally used when referring to increasing the efficiency of a manufacturing system, is a relatively new word and was first coined in 1947 to describe the automatic handling of workpieces. The concept to which it is applied is not new however; in the nineteenth century technological advances led to mechanization of production processes which greatly increased productivity.

The earliest examples of mechanization in engineering were the use of multispindles and power feeds in machine tools. Later developments included sequence-controlled machines which, once set up, could produce large numbers of identical components faster, and at higher rates, than manual machines. These early efforts, however, although greatly increasing productivity and efficiency, were generally purely mechanically operated and worked in isolation from one another.

Around 1940, the next step forward was made with the introduction of transfer lines for the machining of aircraft-engine components. These consisted essentially of a group of drilling, reaming, tapping and milling machines arranged along both sides of a conveyor along which workpieces, clamped onto special holding fixtures, were moved from work station to work station where different operations were performed. All the operations were interlocked so machining could not occur unless all the workpieces, one at each work station, were positioned correctly, and transferring was impossible until all tools and clamps were clear. The transfer-line machines thus performed mechanically both the co-ordination and work-handling functions normally carried out by production control and progress personnel.

Although in-line transfer machining systems developed rapidly and proved to be a tremendous step forward in increasing productivity, they did not entirely overcome one of the biggest problems in all machine shops, that of the actual handling of workpieces. Handling between workstations was avoided but someone still had to load components at the first station and unload them at the last, and work still had to be moved to and from the transfer line. Furthermore, developments in metal-cutting technology brought about the condition in which it was possible for the machine to turn out work faster than a man could load or unload it. This meant that machines were not used to their maximum capacity and handling time was out of all proportion to machining time.

A second problem with the mechanized transfer-line machines was that faults caused by, for example, worn or broken tooling were only detected at some later inspection stage by which time a great deal of scrap may have been produced. The corrective action, replacing the tools and resetting the machine, was performed manually and involved long delays during which the entire line was stopped and no production was possible. The only way to avoid faulty components was to do preventative maintenance on the line and change tools at predetermined regular intervals whether worn or not. Machine tools have now been designed with integral loaders, feeders and unloaders, inspection devices and tool-wear compensation systems. These, combined with improved conveyor systems and industrial robots, have formed the basis of production systems in which it is possible for work to progress from raw stock to finished parts without being touched by hand. Under such conditions it is essential that there is a method of telling whether machines are performing to programme. This is achieved by using feedback control systems to monitor the functions of the production system, which may be described as truly automatic. An automatic system may thus be defined as one which will carry out a pre-set programme or sequence of operations, at the same time measuring and correcting its actual performance in relation to that programme.

By automating work handling, machine tool and information feedback systems, it has been possible to attain very high levels of efficiency in manufacturing. The method of obtaining this efficiency has resulted in very expensive systems, rigidly designed for the production of specific items and only of use in the mass-production sectors of industry.

The problems in batch manufacturing still remain and intensive efforts are being made to apply the techniques of automation to all parts of batch-manufacturing systems from product design to supply of finished parts.

8.2.6 Numerical control

The early automatic manufacturing systems were based on special-purpose machines and work-handling equipment designed and constructed to carry out a single job with little or no variation allowed, with the specific purpose of obtaining high output of accurately made products. Such systems are of little use in batch manufacture.

One of the most significant attempts at applying the technique of automation into batch manufacture was the introduction in the 1950s of numerically controlled (NC) machine tools. These are a general class of automated machine tools which do not rely on orthodox mechanical means for sequencing and positioning functions but are based on more versatile and sophisticated means for programming which depend heavily on applied electronics.

The problem of controlling a machine tool can be divided into the sequencing function, which includes, for example, spindle drive, speed selection, tool selection and the control function, which includes cutting rate and tool position. In a simple sequence-controlled machine, operations may be selected by means of pins inserted into a matrix board, tool positions being fixed by mechanical stops. Such machines, which are usually referred to as 'plugboard autos', although relatively cheap to buy and simple to control, require a longer time for set-up than more sophisticated numerically controlled machines in which a control programme, usually in the form of a punched paper tape, provides the sequencing and positioning commands and servo systems provide the necessary controls. As a result, plugboard-controlled machines tend to be used for batches of more than one hundred and tape-controlled NC machines are most suitable for smaller batches.

The advantages of the NC machine over conventional equipment in batch manufacture include:

(*a*) The ability to produce components of consistent geometry and quality at high rates for long periods so reducing scrap and rework. It is possible that one NC machine may replace three or four conventional machines so reducing the cost of floorspace and manpower.

(*b*) The use of long control programmes and automatic tool changers make it possible to combine many conventional operations into one NC operation and so reduce the amount of work handling and ultimately the lead time of a component and the amount of work in progress.

(*c*) Rather than relying on jigs and fixtures for geometrical information, as do conventional machines, the NC machine obtains all the required information from the control programme. As the cost of modifying the control programme is considerably less than that of modifying jigs etc., NC machines make design changes more easy and cheaper to embody and so increase the flexibility of the manufacturing system.

As NC machines became more sophisticated it was necessary to employ aids to their programming. This led to one of the first applications of computers in manufacturing.

8.2.7 Computer-aided manufacture

The developments in computer technology during the early 1960s led to a rapid expansion in the number of applications of computers. Originally it was thought that they would only be used in scientific applications but, as they became more easily used and more readily available, many applications were found including many in business and industrial manufacturing. About ten years ago, the use of computers had advanced so far that the term Computer-Aided Manufacture (CAM) was coined to describe the application of computers in manufacturing systems. The mass production industries, and in particular the continuous fluid-process industries, such as those involved in petrochemical refining, were able to make use of the capacity of the computer for large-scale data storage and rapid calculation and soon developed computer-based manufacturing systems. More recently, there have been many applications of computers in batch manufacturing. These may be divided into the *off-line processing* of data pertinent to product design and manufacturing planning, and the *on-line control* of production processes and information.

Off-line applications are those in which the computer is remote from the manufacturing system and operates independently of it. Initially, computers found their greatest uses in manufacturing in finance and accounting and in the preparation of NC part-programmes, since these areas have customarily used computational methods and calculating equipment. Later developments have included computer-based systems for production planning and control and for computer-aided design of components.

It is convenient when considering CAM systems to consider planning and control as separate subsystems. The computer-based production planning systems that have been developed typically include program modules and procedures for:

(*a*) requirements planning, which will calculate the required capacity and materials from a sales forecast or order load
(*b*) capacity planning which will roughly calculate a schedule for loading machine tools and other resources with a planning horizon of the order of one year and a planning interval of the order of one month
(*c*) scheduling to provide detailed loading of each resource group in a short period with a planning horizon of the order of one week and a planning interval of the order of a few hours.

Typical computer-aided production control systems involve modules and procedures for:

(*a*) purchase control including replenishment of raw materials, stock control and purchase order control
(*b*) production and assembly control including materials control, load control, inventory control, tool control, job control and dispatching.

The off-line applications are usually performed by what is known as batch processing on a large centralized computer. Data is collected from the plant manually and, together with new production orders and work loads, this serves as input to the planning system which may be run daily, weekly or monthly depending on the production type. The output from the computer generally consists of listings from which the detailed schedule and necessary orders can be extracted. The off-line type of applications are usually the result of applying the computer to the manual procedure. The speed and accuracy of the computer permit the optimization and simulation of plans which would be impossible in the time available using manual means. The discipline of installing a computer system also results in improvements which are due to the fact that procedures and activities are analysed and improved when converting manual procedures into computer logic.

In the early 1960s computer-based systems were developed for the direct digital control of continuous fluid processes. These were examples of *on-line applications* of computers in manufacturing in which the computer is an integral part of the manufacturing system. On-line systems may be divided into two subsystems: those for monitoring and information systems and those for control of manufacturing processes. The purpose of monitoring and information systems is to register and report production data. This may be data concerning active, idle and breakdown times of different machines, inventory transactions or job status and is collected automatically via direct connections to the computer. These systems provide management with up-to-date information regarding the status of the manufacturing resources and so increase the flexibility of the manufacturing system by easing the decision-making processes. There have been many developments in computer-based control systems which include:

(*a*) The sequence-control of a production line, which may involve the knowledge of production data including the number of pieces produced, cycle time, and idle time.

(*b*) Computer Numerical Control (NC) of machines which is numerical control in which the hard-wired conventional control is replaced by a minicomputer programmed to perform the control functions. The power and versatility of the minicomputer and its inherent reliability permits more sophisticated control at a lower cost than is possible using conventional transistor-based logic.

(*c*) Adaptive Control (AC) of machine tools in which the computer is used to measure, for example, cutting forces and speeds and to control the axis motion and spindle speed accordingly so as to maintain the optimum metal removal rate.

(*d*) Direct Numerical Control (DNC) of machine tools which is the connection of several NC machines to a central digital computer for part-programme distribution and storage.

The characteristic feature of on-line systems is that a dedicated computer is used in real-time mode; that is the computer is available at any time on demand to perform its function.

8.2.8 Integrated computer-aided manufacturing systems

Computers are being used in ever-increasing numbers in manufacturing systems but in a somewhat disjointed fashion. New applications have solved particular problems but the overall contribution of the computer has often been less than forecast. It has also become apparent that it is not enough to superimpose computer technology and techniques onto traditional manufacturing systems. The most promising concept for solving the problems of efficiency and flexibility is the Integrated Computer-Aided Manufacturing (ICAM) system. Such a system would be based on work stations, interfaced by automatic handling systems, which have been designed from the 'floor up' to efficiently interface with the digital computer. All aspects of the manufacturing activity including detailed design, specification, manufacturing engineering, materials management, production of parts, assembly, test, warehousing, sales and service, would be controlled by individual modules of computer software, all of which would be linked together in a hierarchical system. ICAM is a total technology which will involve tremendous amounts of software and will be evolved over a fairly long period of time. The development will be evolutionary, however; each module of software will be self-justifying and will perform a useful role within an existing manufacturing system.

8.3 Functional and information models of a typical batch manufacturing system

8.3.1 Structured analysis

The models presented in this section are based on the structured analysis and design technique, SADT, of SofTech. This technique provides a powerful tool for describing the relationship between activities and data in a system. There are two fundamental models used in SADT: firstly, the *functional model* which is concerned principally with activities and, secondly, the *information model* which is concerned principally with data. The typical building block in a functional model is shown in Fig. 8.1 in which the *activity* (a verb) is contained within a box which receives *input data* (a noun) and produces *output data* (a noun). The top entry to the box is a *control* or *constraint*, whilst the bottom entry to the box is a *mechanism*. On the other hand, the typical building block in a data model is shown in Fig. 8.2 in which the *data* (a noun) is contained within a box which receives *input activity* (a verb): this data is then used or consumed by an activity on the right-hand side of the box. As in the previous case, the data in the box can be subjected to both constraints and mechanisms. Functional models and information models are designed to have not less than three boxes and not more than six boxes, and the output of any box can become an input, constraint, or mechanism to any other box or model. The modelling of a complete system is accomplished by a *decomposition* of an uppermost model called AO in the case of a functional model and DO in the case of a data model. If, for example, AO comprised four activity boxes, A1 to A4, then each of these boxes with their associated inputs, outputs, constraints and mechanisms could themselves be expanded to the next level of decomposition. Thus, if activity A2 comprised five identifiable activities then these would be called A21, A22, . . . , A25. This process of decompo-

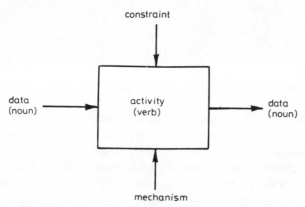

Fig. 8.1   *Basic structure for functional (activity) model*

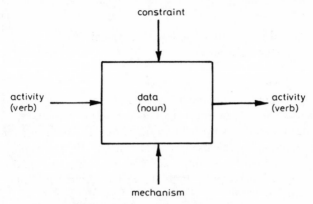

Fig. 8.2 *Basic structure for information (data) model*

sition can be carried on to more and more detailed levels as shown in Fig. 8.3, subject only to the constraint that inputs, outputs, constraints and mechanisms to any box must be maintained at the next level of decomposition: it is possible in some cases to introduce local operations on boxes and these are usually denoted by enclosing them in parentheses. In association with a structured decomposition, it is normal to produce a node list which is a list of names given to each of the activity boxes and/or data boxes.

There are many symbolic notations and drawing procedures which can be used to highlight detail or to simplify the presentation of a model. Two of the more common simplifications are shown in Figs. 8.4 and 8.5. Thus, in Fig. 8.4(*a*), the output of box 1 is an input to box 2, and the output of box 2 is an input to box 1: the mutual exchange of data can be simplified as shown in Fig. 8.4(*b*). Similarly, in Fig. 8.5(*a*), the output of box 1 is a constraint to box 2, and the output of box 2 is a constraint to box 1: this output/constraint relationship can be redrawn as shown in Fig. 8.5(*b*).

A rigorous checking of both the functional and the information models is essential, as they contain complementary descriptions of the same system and must be completely compatible. It should be noted that the models are essentially qualitative and that time is not shown explicitly on the diagrams.

For a full and rigorous description of this modelling technique reference should be made to appropriate literature[1-5] and to SofTech Inc.

8.3.2 Illustrative example

The selected illustrative example is based on a full structured decomposition of a company engaged in batch production of sheet-metal components for the consumer market. The production facilities of this company are organized on a cellular basis rather than on a traditional functional layout. Whilst the company was modelled from the uppermost levels, AO and DO, down to the sixth level of decomposition, diagrams for only the first three levels are

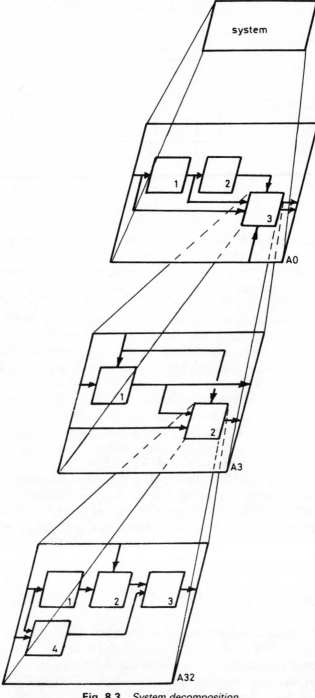

Fig. 8.3 *System decomposition*

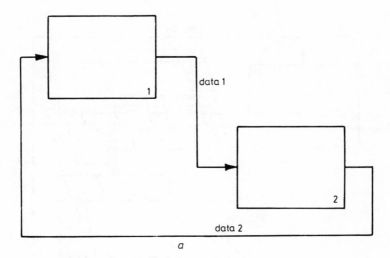

a

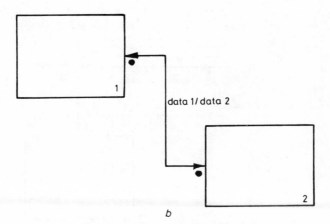

b

Fig. 8.4 *Mutual exchange of data*

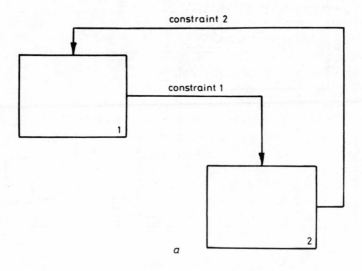

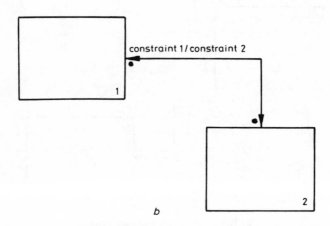

Fig. 8.5   *Mutual exchange of constraints*

Table 8.1 *Activity node list*

AO **RUN COMPANY***
  A1 FORMULATE STRATEGY
  A2 FORMULATE COMPANY PLAN
  A3 IMPLEMENT COMPANY PLAN*
    A31 *MANAGE FINANCE*
    A32 *DESIGN PRODUCTS AND TOOLS*
    A33 *PRODUCE PRODUCTS***
      A331 Prepare production programme*
        A3311 *Plan production methods, times and costs*
        A3312 *Evaluate batch frequency and quantity*
        A3313 *Assess resources and operational characteristics*
            related to company plan
        A3314 *Set production policy and programme**
      A332 Plan and control production*
        A3321 *Explode product*
        A3322 *Determine gross requirements*
        A3323 *Assess inventory levels*
        A3324 *Determine net requirements*
        A3325 *Prepare period work loads**
        A3326 *Schedule in detail*
      A333 Procure commodities*
        A3331 *Vendor selection*
        A3332 *Determine order quantity*
        A3333 *Buy*
        A3334 *Expedite*
      A334 Manufacture*
        A3341 *Make parts*
        A3342 *Stock parts*
        A3343 *Assemble products*
        A3344 *Test products*
        A3345 *Stock products*
      A335 Progress production*
        A3351 *Collect progress data*
        A3352 *Compare plan/progress*
        A3353 *Expedite*
    A34 *MARKET PRODUCTS*

---

* Only these diagrams, which are related to Job Shop Control in Batch Manufacturing, are included in this chapter.

Table 8.2 *Data node list*

DO **RUN COMPANY***
D1 MARKET POSITION, CAPITAL AND RESOURCES
D2 COMPANY OBJECTIVES
D3 COMPANY ORGANIZATION DATA*
   D31 *COMPANY ACCOUNTS*
   D32 *PRODUCT CONCEPTS AND REQUIREMENTS DATA*
   D33 *FACTORY ORGANIZATION AND RESOURCES*
     D331 Production requirements and policy*
       D3311 *Component and product designs*
       D3312 *Annual usage values*
       D3313 *Available company resources and organization*
       D3314 *Arrangement of production requirements*
     D332 Scheduling and control systems data*
       D3321 *Product drawing data and parts lists*
       D3322 *Bill-of-materials and production programmes*
       D3323 *Part stock and work in progress*
       D3324 *Gross requirements and stock list*
       D3325 *Six-weekly work loads*
       D3326 *Two-weekly work loads*
     D333 Bought-out items*
       D3331 *Vendor information, price, delivery*
       D3332 *Order quantity economics*
       D3333 *Orders to vendors*
       D3334 *Delivery performance*
     D334 Process and job-shop data*
       D3341 *Sheet-metal shop*
       D3342 *Stores*
       D3343 *Assembly shop*
       D3344 *Product specification*
       D3345 *Product stores*
     D335 Throughput performance*
       D3351 *Work in progress*
       D3352 *Job status*
       D3353 *List priorities*
   D34 *PRODUCTS*

---

* Only these diagrams, which are related to Job Shop Control in Batch Manufacture, are included in this chapter.

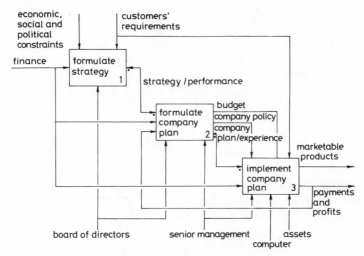

Fig. 8.6   *Activity AO, run company*

included as shown in the node lists presented in Tables 8.1 and 8.2: only those decompositions indicated by asterisks are presented in this chapter. Only the part of the model appropriate to the control of a job shop concerned with the batch manufacture of sheet-metal components is described. This area of the operation is shown as functional models in Figs. 8.6 to 8.13 and as information models in Figs 8.14 to 8.21. The brief notes accompanying these models are as follows:

AO Run company (*Fig. 8.6*). This diagram overviews the operation of the company used for the illustrative example. The *finance* shown as an input to

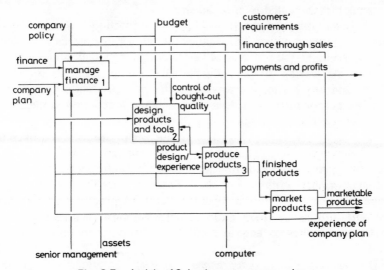

Fig. 8.7   *Activity A3, implement company plan*

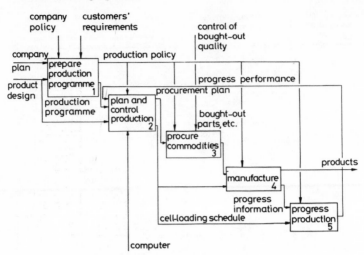

Fig. 8.8   *Activity A33, produce products*

the three activities, A1, A2 and A3, represents borrowing and share capital in the form of additional investment in the company. The *formulate strategy* activity, A1, sets the broad objectives and operational framework for the company against the expectations of the market place and the community. This strategy forms the basis for the *formulate company plan* activity, A2, constructed to meet the company requirements in terms of financial performance. The *implement company plan* activity, A3, is the operation of the company to meet the planned objectives in terms of satisfactions, profits and payments. All these activities are the concern of the company's senior management, as indicated by mechanisms.

A3 Implement company plan (*Fig. 8.7*). Implementation of the company plan is split into four activities. The financial activities operate as a service and control function for the prime activities of design, production and marketing. The payments leaving 'manage finance' are wages, payments to suppliers, dividends and taxes. The company has a standard product range and manufactures no bespoke products. Customer demand is reflected in the design of new products and the redesign or modification of existing products.

A33 Produce products (*Fig. 8.8*). The production of products is broken down into five activities. The preparation of the ·production programme is performed by senior management who produce a twelve-monthly programme and the associated production policy. There is a feedback of performance between this activity and the 'plan and control production' activity. The spares output from manufacture can be single components and subassemblies. A standard computer package is used as an aid to production planning and control.

A331 Prepare production programme (*Fig. 8.9*). The preparation of the production programme contains four activities. The company plan is regularly

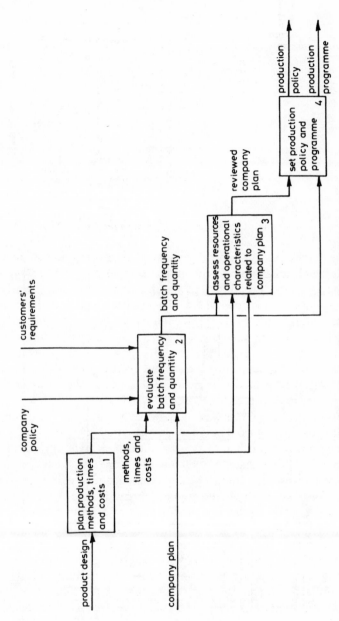

Fig. 8.9 *Activity A331, prepare production programme*

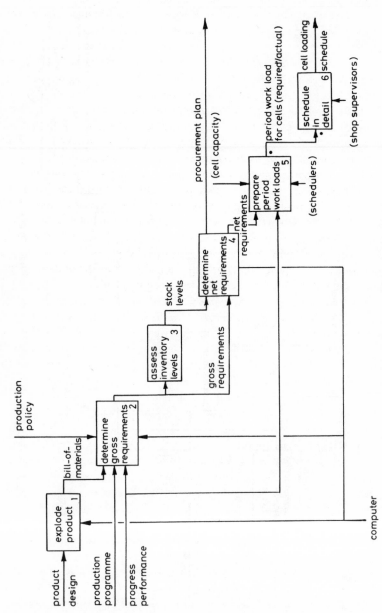

Fig. 8.10 *Activity A332, plan and control production*

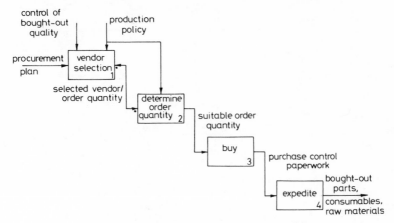

Fig. 8.11 *Activity A333, procure commodities*

reviewed against the background of known methods, times, costs, resources, capacity and operational characteristics.

A332 Plan and control production (Fig. 8.10). The company's computer production control package explodes the product and compares it against the production programme to determine gross requirements. Control of stock is done manually and the data is fed into the computer system for determination of the net requirements. Every four weeks the computer produces a six-weekly forecast of net requirements which is broken down, by a scheduler, into three two-weekly workloads giving details of parts to be made and batch quantities for each machine group or cell. Each two-weekly workload is then scheduled in detail for each cell using conventional bar charts.

A333 Procure commodities (Fig. 8.11). The procurement of commodities is split into four activities. The vendor selection and the determination of the

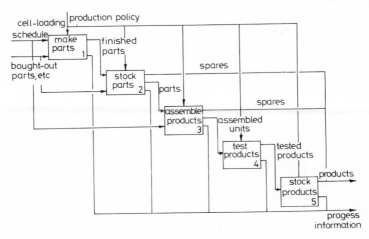

Fig. 8.12 *Activity A334, manufacture*

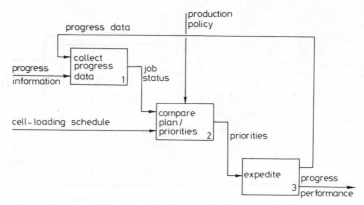

Fig. 8.13   *Activity A335, progress production*

order quantities are interrelated, as each has a bearing on the other. Both activities are constrained by the production policy and the raw material availability. An expediting activity is included for progressing orders.

A334 Manufacture (Fig. 8.12). The manufacture of the products is broken down into five activities, but involves two separate manufacturing installations, one for sheet metal and the other for foundry work. The model is intended to cover only the sheet-metal manufacture, and the foundry manufacturing and control system is not included.

A335 Progress production (Fig. 8.13). Progressing of production involves three activities. The input of progress information is collected from the manufacturing activities described in A334. Actual production is compared with the production plan and shortage, or priority, lists are prepared for submission to each section manager.

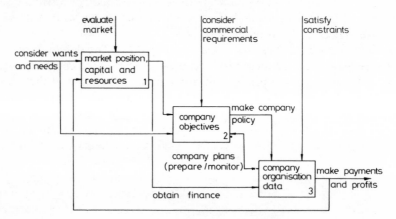

Fig. 8.14   *Data DO, run company*

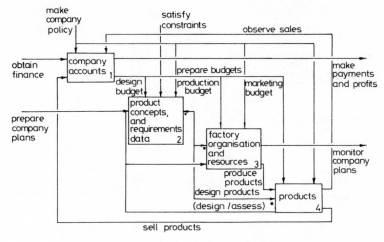

Fig. 8.15   *Data D3, company organization data*

DO Run company (Fig. 8.14). This diagram overviews the information and facilities required to run the company. Based upon the market conditions, the available capital and resources, a corporate strategy is developed. The company objectives are satisfied through the preparation of company plans and policies which are used to drive the corporate organization for the achievement of satisfactions, payments and profits.

D3 Company organization data (Fig. 8.15). In line with company policy, the company accounts are used as the basis for preparation of budgets for design, production and marketing. The interaction shown between boxes 3 and 4 indicates product production with feedback in the form of sales forecast preparation in response to market needs. The interaction between boxes 2 and 3 indicates product design with feedback on product producibility. The interaction between boxes 2 and 4 indicates product design with feedback on the market suitability of product designs.

D33 Factory organization and resources (Fig. 8.16). The design of products and the requirements for these products are the basis of the company organization. The twelve-month production programme is the input to the scheduling and control system which is aided by a computer package for the determination of procurement and manufactured in-house items.

D331 Production requirements and policy (Fig. 8.17). The annual usage values for components are the predicted yearly requirements multiplied by the cost of the component. These annual usage values are used to determine preferred batch quantities for the achievement of minimum work in progress. For example, low usage-value items would be processed through the job shop in large batches at a low frequency (on average, once or twice per year). High usage-value items would be processed in small batches at a higher frequency.

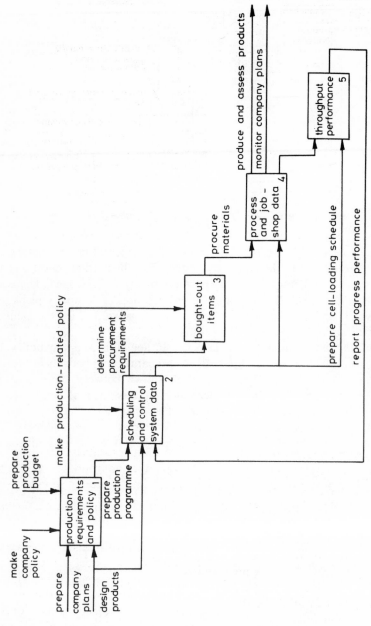

Fig. 8.16 *Data D33, factory organization and resources*

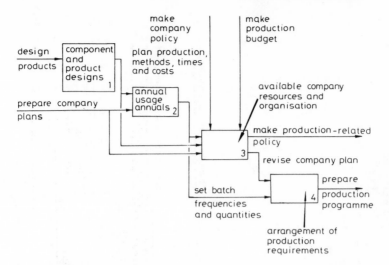

Fig. 8.17 *Data D331, production requirements and policy*

D332 Scheduling and control system (Fig. 8.18). The diagram shows the information required to plan and control production in the sheet-metal shop. The net requirements are determined monthly by the computer package based on stock levels and production requirements. A six-weekly work load is also produced using the computer. The preparation of weekly and daily schedules for the cells is carried out manually.

D333 Bought-out items (Fig. 8.19). The data necessary for purchase of commodities is shown in the diagram.

D334 Process and job-shop data (Fig. 8.20). The facilities required to manufacture components are shown on this diagram. Sheet-metal components are supplied to finished parts stores from the sheet-metal shop.

D335 Throughput performance (Fig. 8.21). Progress data is collected from the sheet-metal shop and compared with the appropriate cell-loading schedule. The problem batches are determined and expedited where feasible, and a report on progress performance is given to production planning and control and to the shop manager.

These models have been decomposed so that employees at all levels in the company can fully understand and appreciate the operations and procedures in which they are intimately involved. In addition, the model can be used as a basis for simplifying and rationalizing the company organization and the control procedures used in the manufacturing process. In particular, this model highlights those areas where the company computer is used only in isolation and that considerable scope exists for the integration of the overall operations of the company using on-line real-time computers.

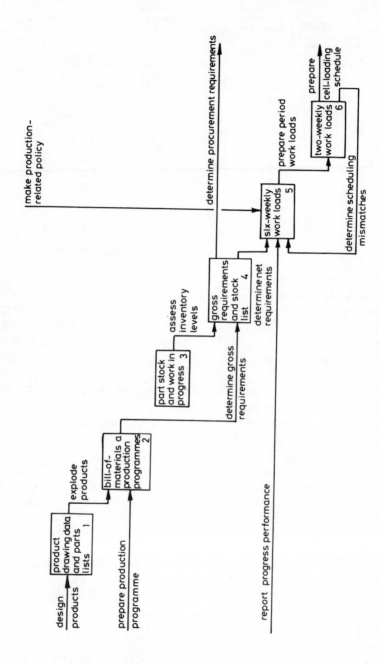

Fig. 8.18   *Data D332, scheduling and control system data*

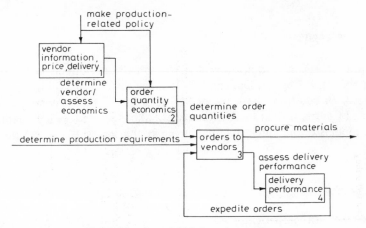

Fig. 8.19   *Data D333, bought-out items*

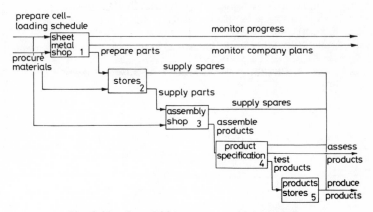

Fig. 8.20   *Data D334, process and job-shop data*

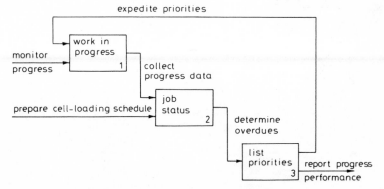

Fig. 8.21   *Data D335, throughput performance*

8.4 Acknowledgments

The author would like to acknowledge the contribution made to the preparation of Section 8.2 by Dr. Duncan McCartney, postgraduate of the University of Salford. In addition, the work of Mr. Ken Swift and his colleagues in the Industrial Centre, Ltd. at the University of Salford formed an invaluable basis for the models of the batch-manufacturing system presented in Section 8.3.

8.5 References

1 ROSS, D. T., GOODENOUGH, J. B., and IRVINE, C. A.: 'Software engineering: processing, principles, and goals', *Computer*, May, 1975, pp. 17–27
2 IRVINE, C. A., and BRACKETT, J. W.: 'Automated software engineering through structured data management, *IEEE Trans. Software Eng.*, SE-3(1), January, 1977
3 ROSS, D. T.: 'Structured analysis (SA): a language for communicating ideas', *IEEE Trans. Software Eng.*, SE-3(1), January, 1977
4 ROSS, D. T., and SCHOMAN, K. E.: 'Structured analysis for requirements definition', *IEEE Trans. Software Eng.*, SE-3(1), January, 1977
5 ROSS, D. T., and BRACKETT, J. W.: 'An approach to structured analysis', *Comp. Decis.*, 8(9), September, 1976

Index